683        $14.95
Wi   Williams, Gene B
          Chilton's guide
    to small appliance
    repair and

DATE DUE

MY 26'92			

CHILTON'S Guide
to
Small Appliance
Repair and Maintenance

CHILTON'S Guide
to
Small Appliance
Repair and Maintenance

Gene B. Williams

Chilton Book Company

Radnor, Pennsylvania

Photos by Gene B. Williams and Deke Barker
Illustrations by John Kwiechen, Vicki Orza, Prue Carrico,
and Adrian J. Ornik, except where source is otherwise noted.

Manufactured in the United States of America

Library of Congress Cataloging-in-Publication Data
Williams, Gene B.
    Chilton's guide to small appliance repair and maintenance.
    Includes index.
    1. Household appliances, Electric—Maintenance and repair—Amateurs' manuals.
I. Chilton Book Company.   II. Title.
TK9901.W475   1986   683'.83   86-47609
ISBN 0-8019-7718-5

SAFETY NOTICE
Proper service and repair procedures are vital to safe, reliable operation of all appliances,
as well as to the personal safety of those performing repairs and using the appliances.
This book gives instructions for safe, effective methods, and the warnings must be heeded. Follow standard safety practices at all times to eliminate the possibility of personal
injury or damage to the appliance.
    It is important to note that techniques, tools, and parts—as well as the skill and experience of the individual performing the work—vary widely. It is not possible to anticipate all the conceivable ways or conditions under which appliances may be serviced, or
to provide cautions as to all possible hazards. Standard and accepted safety precautions
and equipment, along with all local electrical safety codes, must be observed. Heed all
warnings when handling toxic or flammable fluids; wear safety goggles when cutting,
grinding, chiseling, prying, or performing any other process that can cause material removal or projectiles. Some procedures require the use of tools specially designed for a
specific purpose: do not substitute another tool for one designated.
    ASBESTOS WARNING: Some older appliances may contain asbestos insulation.
Asbestos is a known carcinogen; if you encounter it, do not handle it or breathe its dust.
If an asbestos lining is flaking or deteriorating, dispose of the appliance promptly.

    3 4 5 6 7 8 9 0   5 4 3 2 1 0 9 8 7

Contents

CHILTON'S Guide
to
Small Appliance
Repair and Maintenance

Introduction

Small home appliances have often been touted as "labor-saving devices." Salespeople and advertisements usually tell consumers how much time and energy they can save each day by using these handy little helpers. The pitch is that household appliances simplify chores.

Yesterday's luxury item is considered a necessity in modern life. When was the last time you opened a can with one of those old-fashioned hand-operated twist keys? When did you last waste the heat of a full oven to make toast in the morning? (Or try to make toast over a fire?) How often do you wash your clothes in a tub, with water heated on a fire, and use a scrub board to get out the dirt?

Hundreds of household appliances assist us each day in performing tasks more easily, faster, and usually better. Sure, you could manage to live well without them, but would you want to? From the moment a clock-radio or electric alarm awakens you in the morning until you tuck up under a warm electric blanket at night, small electric appliances have become part and parcel of daily living. Some of these devices are designed to perform chores while you sleep, or in your absence. There are automatic timers so that your morning coffee can be perked and ready even as the alarm sounds. As you slept, servo-mechanisms could have watered the lawn and/or garden, recorded a late night movie on a video recorder, recorded phone messages, and could even have taught you subconscious lessons through a pillow speaker.

Morning ablutions in the bathroom often require an electric razor, hair dryers, hot-curlers and automatic toothbrushes or water piks. And breakfast just wouldn't be the same without your toaster, broiler, juicer, microwave oven, electric can opener, or your little handy-dandy personal doughnut maker. All the while, small radios or television sets enter-tain you and issue frequent time checks so you'll know when you have to turn on your phone recorder, use the automatic garage door opener, and take off for work.

Labor-saving devices once considered exclusively for use in the office are now commonplace in private homes, all the way from electric pencil sharpeners and typewriters up to computers. And the home workshop may be supplied with electric drills, sanders, saws, air compressors, and test meters. These tools can be used to fix any of your *other* small appliances whenever *they* malfunction.

The concept of do-it-yourself repair often worries anyone not mechanically inclined. This apprehension might be justified when it comes to repairing an automobile, or a leaking roof. If the work is done improperly, it can be extremely costly to replace those items if damaged by incompetent workmanship.

But small appliances are a different category altogether. What's your first temptation when the toaster won't work? Throw it away! We live in what has been called a "throw away" society. And that doesn't just mean that we throw away a lot of paper, boxes, and uneaten food. We routinely toss out a variety of things because they're not worth the bother to keep.

Small appliances are a good example of this. The cost of repairs by a professional serviceperson is usually as much as — or more than — the cost of a shiny new replacement. Household small appliances generally cost $100 or less, and most carry price tags of from $5 to $75. Professional repair shops today normally have a minimum charge of $30 to $50, just to open up a device to attempt to see what is wrong. So if a $24 electric coffee maker acts up, and the service charge to fix it will be $35, it is cheaper to buy a new one than to hire someone to fix the old. Out goes the old coffee maker, and you plunk down another $24 for a new one.

The fact is that the whole problem with the coffee maker might be a $3 heating element or a 50¢ switch. In fact, the problem might be nothing more than a water carrying tube that has come loose, and needs simply to be pushed back into place (total cost of repair is nothing but a few minutes of your time).

Most of the cost of repairing an appliance is in the labor. It could cost you $35 to fix that $24 coffee maker, even though the actual cost of replacement parts is only a couple of dollars. If the repair shop has a minimum charge (quite common), the actual *cost* of repairing the unit could be zero, with the entire charge being wrapped up in the few minutes of labor needed to take out four screws and push a tube back in place.

The purpose of this book is to save you those labor charges. Why pay someone $40 an hour to take out a couple of screws? Moreover, why toss out your appliance when the repair of most problems is so very easy?

Going back to the coffee maker example — assume that the device no longer perks or no longer keeps the coffee warm. The problem is probably a faulty heating element, which costs just a few dollars to replace. To make the replacement, you usually remove a few screws from the bottom of the coffee maker, slide the bottom cover off, unplug the old element, plug in the new, and put the screws back in. Total time for the job is about 5 minutes. The savings you'll realize is at least $20. Twenty dollars for 5 minutes of work is a pretty good rate of return — in fact, it works out to be $240 per hour! This should make you stop and think about what you're tossing into the garbage can.

The important thing to keep in mind is that for the majority of small appliance malfunctions, you *can* make a successful repair, even if you don't have a mechanical background. It's not as difficult as you might think. Diagnosis of a problem is usually easy. When it's not, a few simple tests will often make it obvious. And you don't need a bench full of fancy equipment and tools to get the job done. Nor do you need years of special training. Try it once and you'll realize just how easy it is. The more you do, the easier it gets — and the more you save. (You might even begin to wonder why in the world we've become a "throw-away society".)

There will be times, however, when your attempts at repair will be useless. You might find that the repair is either impossible, or requires more time, effort and actual cost (for parts) than the price of a new device. Even then you don't lose out.

First, you can use these ailing automatic appliances as a sort of training step for learning more about do-it-yourself projects. The next time something needs fixing, you'll be better prepared to handle it.

Second, and almost as important, in taking apart the malfunctioning device you will have provided yourself with a "spare parts" supply from the old unit. Those can be of obvious value if you buy a new one of the same make and model. What many people don't realize is that quite often the salvaged parts can be used in other small appliances. (Electric cords are an excellent example of this universal adapability.)

Beyond just saving money, there are other reasons for attempting to fix a broken appliance yourself. The item may have been a gift that you treasure. It may not be worth the price of professional repair, but if at all possible you'd still like to keep it around and operational.

In some instances, a favored appliance may be an old model that suits you just fine. You prefer the way the old one did the job and the "quality of workmanship of days gone by." You'd much rather have the familiar appliance back in good working order than have to make do with a newer version you dislike.

In almost all cases, small appliance repair is a case of nothing ventured, nothing gained. And you are in a strictly no loss situation, since the loss is already there if the appliance can't be whipped back into shape through your own efforts. The most it can cost you is some minutes of your time to find out if you can successfully make the repair.

With a few simple hand tools, common sense, and this book, you'll be able to diagnose and fix most of the common problems that can interrupt the smooth and reliable operation of these everyday conveniences. You'll also learn how to carry out some simple maintenance steps to reduce the number of times that repair becomes necessary.

In short, the cost of this book could be one of the more valuable investments you've ever made! You can save its cost many times over with your first job.

In Section One you will be given all the basics you need to successfully repair and maintain the small appliances in and around your home. Section Two is a step-by-step guide to repairing dozens of appliances, arranged in alphabetical order from blankets and coffee makers to telephones and trash compactors.

Chapter 1 describes the basic tools needed for small repair jobs and how to use them. It also lists some of the supplies you might wish to keep on hand if you intend to become a serious do-it-yourselfer.

Chapter 2 is dedicated to the subject of safety. *Nothing* is more important than safety. The chapter is fairly short and to the point, so read it, and then read it again. It's the single most important chapter in the book. Every year people lose their lives because they consider safe procedures to be too boring to learn. Repeat: there is *nothing* more important than safety.

Since almost all appliances are electrically operated, Chapter 3 is devoted to a description of home electrical codes and practices and to demonstrations of how best to handle electrical repairs. It is filled with practical information to help you avoid making costly and possibly dangerous mistakes. Put this information to use and you are highly unlikely to run into trouble while working on any electrical, motor-driven, or mechanical device.

Chapter 4 describes the diagnosis and repair principles for all small appliances. These fundamental procedures will let you handle all but the most complex repairs on any device. This chapter will be of help even if your appliance is not individually listed in Section Two.

Electric motors are covered completely in Chapter 5. These are the power sources for most small appliances. You'll discover that even though many of the small electric motors used are sealed, some can be opened and brought back to life again rather easily. (You'll also find that many times a motor problem is nothing more than a loose connector.) Inexpensive sources for replacement motors are listed, in the event you are unable to get the original one working again.

Chapter 6 covers all the heating elements used in appliances such as blankets, hair dryers, broilers, coffee makers, or any other device where providing heat is one of its functions. You will be shown how to isolate the malfunction to a particular part, how to obtain replacements, and then how to install the replacement properly.

Chapter 7 tells how to maintain appliances to help reduce the number of times that repairs are needed. Sound maintenance practice can go a long way toward keeping your appliances in good working order.

In Section Two, you will be guided step-by-step to troubleshooting and repair of individual appliances. To make finding the item easier, they are presented in alphabetical order. For each appliance there is a discussion of the basic design and operation, a review of common breakdowns, and troubleshooting tips to assist in problem identification and diagnosis. Specific information is then provided for replacement and/or repair of the defect.

All of the most common appliances are represented. If the appliance that is giving you trouble isn't specifically listed, refer first to Section One. Chances are, you'll find exactly what you need to know here. You can also look up a similar appliance in Section Two for more specific information. (If it's a large appliance that is giving you trouble, be sure to see the companion book, *Chilton's Guide to Large Appliance Repair and Maintenance*.)

The First Steps

1. *Read the entire book thoroughly, particularly Section One.*
2. *Study Chapter 2 once again (for safety tips) before you begin. Read Chapter 3 if applicable.*
3. *Perform the basic diagnostics steps of Chapter 4.*
4. *Read the applicable repair description in Section Two.*
5. *Complete the tests on your unit to find the defective part.*
6. *Repair or replace that bad part.*
7. *Use your appliance!*

Before anything else, your first step should be to read through this book completely. Section One is of special importance since it provides all the basic information you'll need for the job at hand, and also gives you the safety tips that will protect both you and the device on which you are working.

Many people will be tempted to skip Section One. Don't. The information in Section One is too important to ignore.

Even more people will ignore Section Two except for those parts that are of immediate importance. That, too, is a mistake. Even if you're working on a coffee maker, the hints on diagnosis and repair of an electric lawn clipper can be handy. Imagine bringing that malfunctioning appliance to a repair shop. Who would your prefer to handle the job — someone who is comfortable only with coffee makers or someone who has a solid overall knowledge of how small appliances work in general? If that person has an overall knowledge of all appliances he or she is more likely to carry out the needed diagnosis and repair successfully. If that person has never worked on anything but coffee makers, he or she will be less prepared to handle the unexpected.

After reading the book, study Chapter 2 again. The importance of safety can hardly be stressed enough. If you can't understand how important safety is and can't spare the time to study Chapter 2 before beginning, toss out the malfunctioning appliance and buy a new one. You'll save money (because you can't be bothered to learn proper repair procedures), and possibly your life.

Steps 3, 4 and 5 on the list are actually more like three parts of a single step. Carry out the diagnostic steps and read the appropriate part in Section Two with its specific diagnostics, tests, and hints for that particular appliance. Once this has been done you'll know what is causing the trouble and you can proceed with the repair or replacement of the defective part.

This brings you to the nicest part of it. You can put away your checkbook and stop watching the ads for appliance sales. The one you were about to toss in the trash will be functioning again.

Section One

Before You Begin

Chapter 1
The Tools and Workshop

A few hundred years ago, the average person was capable of making the needed repairs to the simple machinery used around the home or farm. Despite how complex machinery has become, the same holds true today. The only real difference is that many people don't realize that *they can do it*.

As more sophisticated equipment moves into our private homes, it's easy to think that the task of repair requires more knowledge and experience than the average owner has. As much as that owner wants to avoid the high cost of repair, he or she just doesn't feel capable of handling the job. It seems too complicated for the homeowner who rarely handles a screwdriver let alone the insides of a broiler oven or toaster. The owner with this common attitude may spend hundreds of dollars to have a professional come out to make the repairs, often when the owner can easily do them. (You'll find that the majority of repairs are very simple. Have you ever known someone who paid $50 to have a repair technician come out just to push in the plug? It happens all the time.)

Repairs to appliances are not as difficult as you might think. Once you've overcome that first major hurdle of realizing that you *can* get the job done as well as—and often better than—that professional, you've already accomplished at least 75% of the job.

The goal of this book is to give you the remaining 25%. It will show you some of the steps that professional technicians use to test appliances, and to quickly determine what part has malfunctioned.

GENERAL PURPOSE TOOLS

Before you attempt any repair work, it is necessary to acquire a few basic hand tools. You probably already have most of these in your toolbox. If you don't, it's past time that you made the investment. These things are too useful to worry about the few dollars they will cost. It's a one-time investment if you buy wisely. It's also an investment that will pay off many times over.

As a simple example, imagine that the outlet for your hot plate has become faulty. You could pay a professional $50 or more (plus parts, which are charged to you after a hefty markup has been applied) to replace it. Or you could dig out that $2 screwdriver and do it yourself. That's a savings of $48—and you get to keep the screwdriver.

Basic Tools

Screwdrivers
(blade and Phillips)

Wrenches

Socket set

Allen wrenches

Nut drivers

Pliers

Needlenose pliers

Wire cutters

Knife

Soldering tools

Multimeter
(volt-ohm-milliammeter)

Appliances are held together with various types of screws, nuts, and bolts, along with other fasteners. Once inside, you'll also find that many of the components are held in place in much the same way. Before you can do any servicing, you'll have to acquire the tools needed to get inside and to work with the components.

You will need a few standard screwdrivers with tips of various sizes. All must have insulated handles, since most appliances are electrically operated. Different sizes of screwdrivers are required because appliances use different sizes of screws. Using the wrong size screwdriver can cause several problems. First, it may not work at all. Second, it may damage the screw, making it extremely difficult or impossible to remove even if you get smart and switch to the proper size screwdriver for the job. Third, you can damage the appliance. Fourth, and most important, you can injure yourself.

Use the right screwdriver for the job. The few extra dollars you'll spend for a complete screwdriver set will pay off.

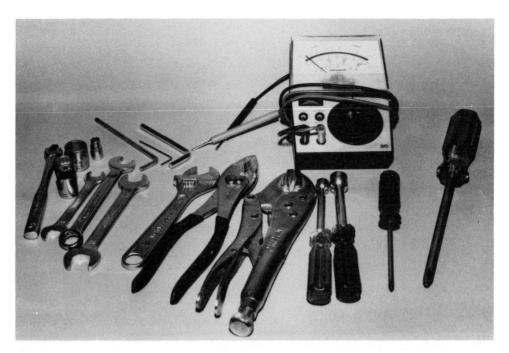

FIG. 1-1 Basic tools needed. From left, socket set, wrenches, allen wrenches, crescent wrench, pliers, vise grip, nut drivers, screwdrivers, and VOM.

BLADE   PHILLIPS   HEX   TORX

FIG. 1-2 Screwhead types.

Another type of screw head is called a Phillips, which is like a four-sided star. You'll need at least one with a fairly small head. It's also handy to have a Phillips screwdriver with a larger head on it. A few large appliances require a Phillips screwdriver with a very small head. Screws of this type are sometimes used to hold smaller components in place. Be sure that the screwdriver has an insulated handle. And be sure that you use the right size screwdriver for the screw.

Somewhat like a Phillips head is a relatively new design called a Torx. This kind of screw head is highly resistant to stripping. Almost certainly more and more manufacturers will begin to use this kind of screw. If your appliance uses this kind of screwhead, *do not* try to remove it with anything other than a Torx driver. Using a Phillips or blade screwdriver will damage the screw, the driver, and possibly the appliance.

Occasionally you'll come across other types of screw or bolt heads.

FIG. 1-3 This kind of fastener gives you a choice of tools. Whenever possible, use a nut-driver instead of a screwdriver.

The most common of these have a hex head, either indented (use an allen wrench) or with a nut-like head (use a wrench or nutdriver). In many of these cases, the screw or bolt will have a head that gives you a choice of tools. Commonly, the head will have a slot for a screwdriver blade, surrounded by a hexagonal nut that can be turned with a wrench, socket, or nutdriver. When given this choice, it's always preferable to use the nut head instead of the slot. The job is easier and safer this way.

A set of wrenches will often be needed to get inside the appliance, and sometimes a wrench will be the only way to take care of certain jobs. For example, replacing a belt often requires loosening and tightening the motor mount bolts. If you don't tighten the bolt completely, the belt will soon be flopping again.

An alternative to wrenches (but not necessarily a replacement) is a socket set. These days you can find inexpensive sets in many places. A complete 40-piece set might cost as little as $5. These tools are obviously of lower quality, but are certainly sufficient for the occasional handyman.

Nut drivers have been mentioned several times. If you intend to do much repair work in your shop, another useful acquisition is a set of nut drivers. These are like fixed socket wrenches, or like a combination of a screwdriver and a socket.

A pair of pliers with plastic insulated handles and a pair of needle-nosed pliers will be of help to you. These are used for gripping. *Do not* use them in place of a proper wrench or socket.

As always, it's best to use the tool meant for the job. A separate wire clippers is generally better than the one built into a pliers. Although you can use a sharp knife to strip away the insulation from a wire, a wire strippers will do the job better, quicker, and more safely. (Even so, a sharp pocket knife is something no tool kit should be without.)

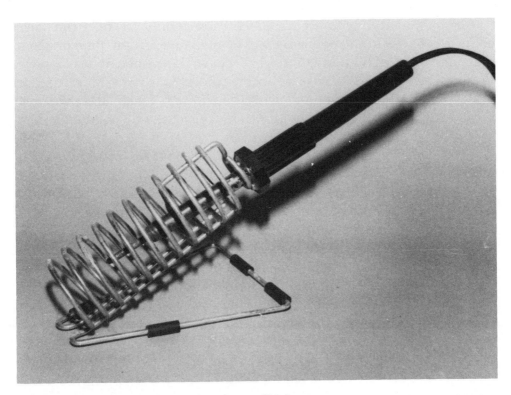

FIG. 1-4 This soldering tool is fine for small jobs.

Another convenient combination of tools for the home workshop is the small soldering iron and a desoldering tool (often called a "solder sucker").

For some appliance work, a heavy soldering iron or gun of 100 watts is best. This type of iron is too hot for working on sensitive solid-state electronic circuits such as the transistors and IC chips in a television set, but the extra heat is often needed for effective soldering work on many home appliances.

When buying a soldering iron or solder, tell the person at the counter what type of work you will be using it for. Most shop personnel are familiar with the correct style and model you should be using.

The solder sucker, which literally sucks up excess solder from a heated joint, is used to remove old solder when you are attempting to replace a part or disconnect a wire so that meter readings may be taken properly. These tools come in several different types, from a simple squeeze bulb to a fancy spring-loaded heavy-duty model. If you have to disconnect a soldered part, first you must melt the old solder joint and, while the joint is still hot, use the desoldering tool to remove the melted solder. The component or wire can then be removed from its original position easily.

Before attempting to use a soldering iron, study carefully the pages on its proper use in Chapter 3. Better yet, read this section, then practice, practice, practice! Soldering is fairly easy, but it must be done correctly. In electrical work, always use resin core solder. Acid core solder, used in some forms of construction requiring the binding of metals, is corrosive. If used for making electrical connections, the joint will eventually deteriorate and the electrical contact will be destroyed.

One tool you can hardly afford to be without—a VOM (volt-ohm-milliameter)—is discussed at the end of this chapter. For the moment, jot this down as one of the "must get" tools, and if you aren't already familiar with its operation (and even if you are), turn to the last section in this chapter for more information.

As you become more sophisticated in repair and maintenance work, you may wish to obtain other tools, such as a small electric drill, and possibly a set of clamps, vises, and holders in order to work on jobs requiring a "third hand." A third hand is a tool made up of alligator clips and arms built into a stand; it aids in holding wires and other small objects that need soldering or other attention.

Heating elements in particular can't be soldered. While in operation they get too hot, and regular solder would simply melt off. Consequently, you may need a good crimping tool, or possibly a small torch and silver solder. More information on how these are used is contained in Chapter 5, "Heating Elements."

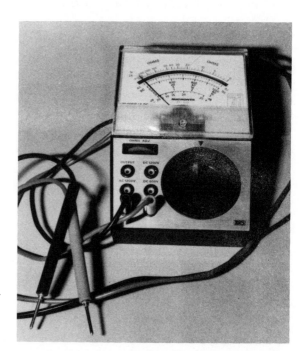

FIG. 1-5 The VOM is one of the most important tools for the do-it-yourselfer. You don't have to spend a lot of money to get one that works.

SUPPLIES

You don't need a huge stock of supplies to carry out successful repairs. As mentioned in the Introduction, you can slowly build up your collection of spare parts as you go along by scavenging unfixable appliances. Meanwhile, the local parts supply store will serve very well as an out-of-home stockroom.

Your only real need is for supplies that will be used for many different repairs. Several of these are described below.

Handy Supplies

Electrical tape

Denatured alcohol, cotton pads and swabs

Spray lubricant

Connectors of various types

Wire, 10 to 16 gauge

A roll of good quality electrical tape can be indispensible around the home. This is the plastic-coated insulating tape, not what is often called "friction tape." If you don't already make a habit of having plenty of electrical tape around the home, once you get some you'll wonder how you got along without it.

For cleaning and preventive maintenance of appliances, the do-it-yourselfer should have on hand a bottle of denatured alcohol (not rubbing alcohol) and cotton pads and swabs. The more pure the alcohol is, the better. You can find it on the shelves in most drugstores with a purity up to about 90%. This is fine for most jobs. Even so, you may wish to go to the trouble of getting a better grade, such as 99%. The difference in cost is very small and assures that you won't be contaminating when your goal is cleaning.

A can of spray lubricant, such as WD-40, and a supply of light machine oil can help in preventing costly appliance breakdowns if used properly. Used improperly both can cause more problems than they'll cure. The trick is to use as little lubricant as possible, and to thoroughly clean away any excess afterwards.

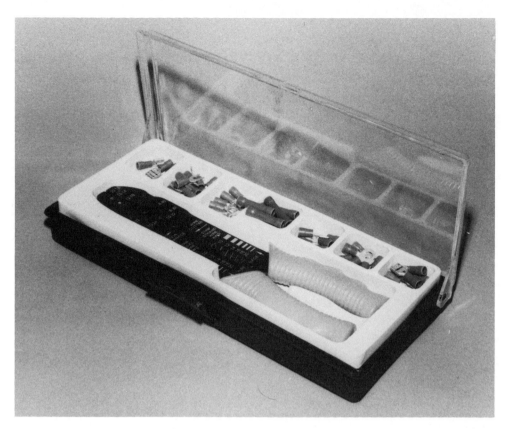

FIG. 1-6 This kit comes with a variety of different kinds of connectors and a multi-purpose tool.

Also desirable is a supply of various connectors.

You should be able to find connector sets at most hardware and electronic stores. These contain several each of different kinds of connectors and are a good way to get started. Later on you can get packages of specific connectors. (It's almost always worth getting a package rather than single connectors, unless you need a specific one.)

For about $10 you can buy a more complete set with connectors and a wire stripper/crimper/cutter tool. For occasional jobs, these combination tools work just fine. If you're planning to do a number of repairs, it's worth it to pay a little extra for a higher-quality set.

A supply of different sizes of wire is also useful. Not just any wire will do. You can often (but not always) replace a smaller gauge wire with a wire of larger gauge. The reverse isn't true.

The wire that carries a tiny control voltage might be just a few strands of thin copper covered by a light coating of insulation. The wires going to the heating elements of an electric range are probably heavy wire with matching insulation. If you try to replace the heater element wire with the small one for a control, the best that will happen is that the wire will melt and everything will stop. Hopefully nobody would be silly enough to try such a thing. However, it's all too common for someone to replace a heavy 10-gauge wire with 14-gauge wire because that happens to be at hand. Failure in this case won't be as quick, but that makes it all the more dangerous. That too small wire will be getting very hot, possibly hot enough to melt away the protective insulation. And all of a sudden you have a deadly 240 volts just waiting for you to touch the wire. The voltage may even be conducted through the metal body of the appliance.

A well-equipped work bench will have small spools of wire from gauges 10 to 16. If buying a spool each of the different wire sizes is more than you care to spend, don't worry. Many hardware stores carry large spools of wire and allow the customers to buy whatever length is needed. It is simply more convenient to have a supply at home.

The best way is to get things as you need them, but in slightly larger quantities than are required for the job at hand. For example, if you have a need for 16 inches of 10-gauge wire, get an extra few of feet at the same time. The additional wire will cost you very little. It allows you to make mistakes, and if you don't make mistakes, it gives you some extra wire for the next time that size wire is needed. You might also find that same gauge wire in a small spool of 25 or 50 feet for about the same cost as the few feet you actually need.

Once you get going, you'll find a hundred uses for electrical tape so you can safely buy that "Special Today!" package of 10 rolls. On the other hand, if you have to replace the thermostat on your hot plate or fry skillet, chances are good that you'll never have to do so again. There's no need to buy two.

THE VOLT-OHM METER (VOM)

To test for electrical voltages, measure component values, and check for continuity (continuous wiring contact from one point to another), you will need a multimeter (volt-ohm meter). Testing a wall outlet *can* be done easily with a tester made specially for this purpose, and some technicians prefer this simpler tester. Others feel that the only viable way to test *any* voltage is with a meter.

Fig. 1-5 is a photograph of a simple volt-ohm-milliammeter, or VOM. Today, some pretty fancy digital VOMs are available from hardware and electronic supply stores for less than $50. Simple models with a needle and scales are available at prices less than $15.

Although a more expensive meter is usually better, you can get by very well with one of the very simple meters. Almost all meters made today, including those $10 specials, are accurate enough for your purposes.

Whichever VOM you get, it should be able to measure at least voltage and resistance. Unless the meter has built-in automatic switching, the meter should have several ranges for each kind of reading.

The voltages you will be measuring will be somewhere between 5 to 50 volt dc (vdc), and 120 or 240 volt ac (vac). The meter you choose doesn't necessarily have to provide exactly those settings. For example, I have one meter that offers dc voltage choices of 0.6, 3, 15, 60 and 300+ for a setting. There is no specific setting for 12 vdc, but this doesn't matter. By putting the meter in the 15 vdc range, 12 vdc can be measured very accurately.

The meter you purchase should also be capable of measuring resistance (in ohms—the "O" of VOM). To many newcomers, reading ohms is useful only to check the value of resistors. This is actually the least useful way to use a VOM. (Resistors are already color-coded as to value anyway.) Resistance tests are useful for checking such things as continuity (a complete and continuous electronic path).

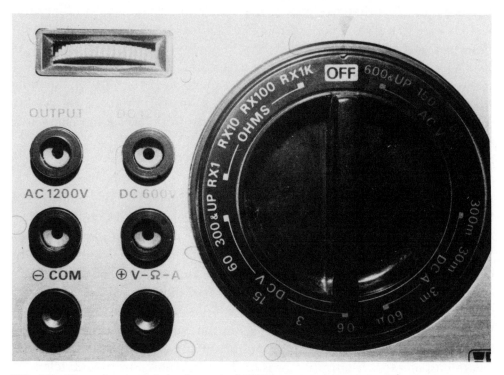

FIG. 1-7 The selector switch of a typical VOM.

A third function on many VOMs is current reading. An analogy often used is one that compares electricity in a wire or circuit to water in a hose. In this analogy, current would be like the number of gallons per minute of water, with voltage being the force or pressure of the water. The milliameter function of a VOM measures the quantity of electricity in thousandths of an amp. It's very rare to need this function when testing or repairing appliances. Usually, if you need to measure current flow, that flow will be greater than 1 amp and could be as high as 50 amps. A special meter is needed for this.

Carefully read the instruction manual that came with your VOM. The operation of each model and style of meter is slightly different. Also, meter scales read differently, so be sure to study your meter's scale or digital readouts so that you can interpet the meter or dial readings the instrument gives.

A VOM can be damaged internally if it is set to read resistance (the ohms scale) and then the probes are connected to a voltage source. Be sure you know what you want to measure (volts, ohms, current) *before* connecting the probes or turning on any power source. It can also be damaged if you use the correct setting by the incorrect range (set to read 3 vdc while probing a 60 vdc circuit).

The following sections will give you the basics of using a VOM. If you are not familiar with its use, be sure to practice with the meter and learn its functions well before probing inside one of your appliances. Take readings of the various outlets in your home to check the ac voltage scales. Test old batteries to get used to the dc scales. Use unplugged extension cords to get practice in testing for continuity — to see if the wire is broken (or "open").

In short, get used to the meter before you have a real need to use it.

TESTING FOR VOLTAGE

As you'll learn in this book, finding the cause of a malfunction (diagnostics) is nothing more than a process of elimination. There are only so many reasons why something fails to work as it should.

Assume that an appliance has failed completely. There are just four places that the problem could be. It's possible that there is no power coming into your house; or that the trouble is between the house service box and the outlet; or it might be getting that far but can't get into the device due to a faulty power cord; or it could even be something in the appliance itself.

The VOM will help you to track it down quickly.

Before you even get out the meter, eliminate the obvious, such as the

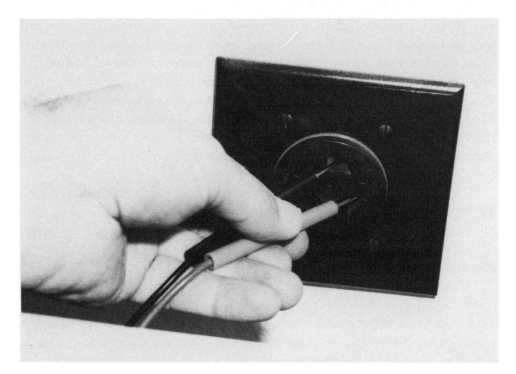

FIG. 1-8 Testing a wall outlet for power.

appliance being unplugged. If it *is* plugged in (are you *sure?*) the VOM can be used to test for voltage at the outlet.

It's possible to test the outlet by plugging in a lamp or another appliance. However, there are times when the incoming voltage is present, but not high enough in value to operate that particular appliance. A lamp, for example, might look normal, when in fact the incoming voltage is insufficient for your device. Testing with a VOM will tell you for sure. If your area is prone to "brown-outs," using a meter to test the outlet is especially important.

To test a wall outlet, set the meter to the correct ac voltage range (120 vac for a standard wall outlet, and 240 vac for heavier appliances). Keep in mind that you are probing potentially deadly voltage. Hold the probes by the insulated handles only! Do not touch the metal of the probes.

Very carefully insert the two probes into the flat slots of the outlet. The meter should read 117 volts on the scale, or very close to it. With a grounded outlet (two flat slots and a round one), you should be able to get the same reading between the ground (the round hole) and the slot with the incoming line, but not between the ground and the other slot. (If you get a reading between the ground hole and both flat slots, it's time to call in an electrician.)

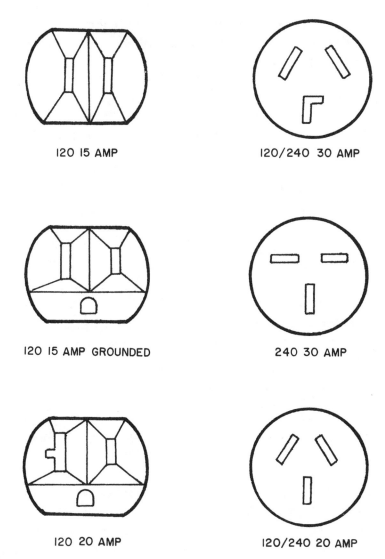

120 15 AMP

120/240 30 AMP

120 15 AMP GROUNDED

240 30 AMP

120 20 AMP

120/240 20 AMP

FIG. 1-9 Typical outlets; standard 120 vac wall outlets and 240 vac outlets.

With a 240 volt outlet, set the meter to read in the correct range and probe the slots in the same way. If everything is normal, you'll get a reading of 117 volts between the lower slot and each of the two upper slots, and a reading of about 234 volts between the two upper slots.

If you get the proper readings at the outlet, you've just eliminated two of the four possibilities. In fact, even if you *don't* get the proper readings — assuming you've used the meter correctly — two of the four have been eliminated. If the readings are correct, then power is getting to the outlet and the problem is either in the cord or in the appliance. If the readings aren't correct, then the appliance is probably fine (or it has a short circuit that is causing the fuse or circuit breaker to blow). The problem is either with the power company, or someplace between the service box and the outlet.

In a few cases, an appliance will have a built-in power supply that takes the incoming 120 vac and converts it to other values of ac, or to various values of dc. Quite often these are marked on the appliance. Measuring is the same as measuring voltage across any wires. Keep in mind that electricity has to have a complete path in order to work. It has to come in somewhere, go through the circuit, and then exit in a complete path back to the supply.

When measuring dc voltage or current, the polarity of the test leads is important. Direct current (dc) is current moving from point A to point B in a circuit in a single direction. Theory assumes that current moves from source (the "hot" side of a circuit) to ground. The common or ground lead of the meter goes to the ground side of the circuit, the red probe goes to the hot side to read dc voltage or dc current. Check, and then recheck, the polarity of the test leads connecting voltage to a circuit before applying power. If hooked up backwards, a meter with a moving needle will have the needle deflect downward, in a reverse direction. Digital meters will display a − (negative) value. To get correct readings, just reverse the polarity of the probes.

Check (and then recheck) the *range* of your meter before applying power. You can damage the VOM by trying to read voltage from a supposed 50-vdc source when your meter selector switch is set to the dc 0-5 volt scale.

It is not necessary to observe polarity when measuring ac voltages. Alternating current moves back and forth in a wire or circuit, first in one direction and then the other. Like resistance, ac voltage and current can be checked without worrying about which probe is on the negative and which is on the positive terminal of the power source.

When testing ac voltage, be sure that you locate and identify the specific ac connections before probing. For dc, it's generally not as critical. You can often safely touch the black probe to any known ground and then use the red probe to take the readings. *Never* touch the black probe to the chassis (and the red to the suspected spot) to test for ac voltage. The meter becomes a part of a complete circuit in this case and can make the entire appliance deadly to touch. In an appliance where both ac and dc are present, be careful that you know which wires and connectors carry which voltage, and at what value, before probing.

TESTING FOR CONTINUITY AND RESISTANCE

If there is power at the outlet but the appliance is dead, the trouble could be a damaged power cord. It is sometimes possible to probe the connectors where the ac voltage enters the appliance. Even so, it's best to test the cord and connectors by checking for continuity.

As mentioned in the section above, for electricity to work, it must have a complete and continuous path. If that path is broken anywhere, current stops flowing. In a power cord, if one of the two wires is broken or has come loose from the connector, the pathway is no longer complete, and there is no continuity.

The resistance (ohms) setting of your VOM will let you test for continuity. In simple terms, if the wire being tested is intact, you'll get a reading very near zero ohms (no resistance to current flow). For a meter with a needle, this will show by the needle swinging full-scale all the way across the meter. If there is a break in the wire, the needle won't move, indicating that the resistance is infinite (no continuous path, and no continuity).

Testing for continuity is one of the best diagnostic tests you can make in appliance repair. First be sure that the appliance being tested is turned off and unplugged. If you're testing the power cord, you'll have to be able to access both sides of the cord. This means that if the cord

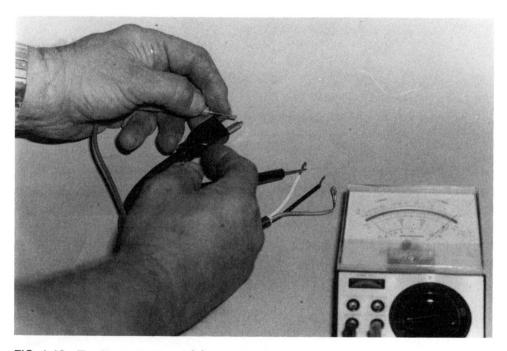

FIG. 1-10   Testing a power cord for continuity.

doesn't unplug from the appliance itself, you'll have to open the appliance to get at the connectors inside.

Put the selector switch of your VOM into the resistance (ohms) range. The actual range isn't critical, but it is generally best to use the lowest setting possible. (This is usually the X1 range.) Touch the ends of the two meter leads together. The meter should indicate zero ohms, or no resistance. The meter is showing that there is an unbroken, continuous path from one probe to the other. If there is a slight reading, use the "zero adjust" control on the meter to set for zero.

Now touch one of the probes to one side of the wire and the other probe to the other side of the same wire. You should get a reading of zero ohms.

To test for a short circuit along the cord, touch one probe to one of the two wires, and the other probe to the other wire. A reading of infinity—which is what you want—shows you that there is no path between the two legs. Any reading at all shows that current can flow between the two wires—and it shouldn't.

In sequence, touch the other probe to each of the three prongs on the plug. *One* of the prongs should show continuity (zero ohms) to the wire

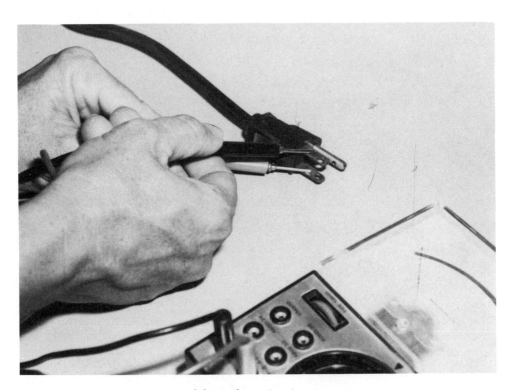

FIG. 1-11   Testing a power cord for a short circuit.

at the other end, the others should show open (a reading of infinity, and no deflection on the meter).

This is also one swift way to trace a wire from one point to another, to see which wire is which in a circuit. When you read continuity, you know those two points are wired together someplace or other in the appliance. If your meter indicates "open" (no connection) from any of the three prongs to the other two, you have no short circuits in the cord.

In turn, move the probe at the open end of the cord to the other two wires, and make sure that each shows continuity (zero ohms) to *one* prong, and open (infinite resistance) to the other two. The same type of meter test can be made with one of the simpler push-on type of clamping replacement plugs, if the end of the cord away from the plug is also exposed. If it's a two wire plug, the VOM must indicate that you have continuity from one prong to one wire, and an open circuit from that same prong to the other wire. It must also indicate continuity from the second prong to the second wire, and an open from the second prong to the first wire.

Testing for resistance is an excellent way to find out if a particular component or element is starting to give out. A heater element, for example, can be tested easily with a meter. With the power off and the appliance unplugged, disconnect one side of the element. Set your VOM to read resistance and touch the probes to the two ends of the element. You should get a reading near zero ohms. If there is no meter deflection, the

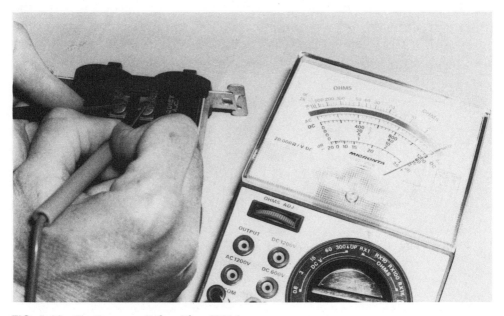

FIG. 1-12   Testing a switch with a VOM.

element has become "open" and there is a break in the wire. A reading that is not infinite (no deflection) but shows a high resistance is a sign that the element is starting to wear out. Either way, it's time to replace that element.

This same kind of testing can be used elsewhere in the appliance. (Always be sure that the power is off and that the appliance is unplugged before you begin.) By touching the probes to two ends of a wire, you can determine if a complete path exists (continuity) or if there is a break in that circuit. It can also be used to test for short circuits by touching one probe to the chassis of the appliance and the other to the suspected component.

Even switches can be tested with the VOM. Find the connectors across the switch and touch these with the probes (again with the power off, the appliance unplugged, and the meter set to read ohms). With the switch in the off position, there should be no reading; and with it switched on, there should be a reading of zero ohms. If this doesn't happen, the switch is bad.

TESTING FOR CURRENT

Current is the amount of electricity in amps flowing through a circuit in order to make it work. To measure current the meter must become an actual part of the circuit. You can measure resistance or voltage by placing the meter across (in parallel) with the circuit, with one probe touching one end of a wire or on one terminal and the other probe touching the other end of the wire or a second terminal. Current can only be metered from within a circuit.

To give a practical demonstration, take a working circuit consisting

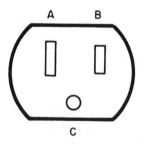 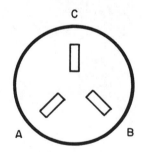

FIG. 1-13   On a 120-volt ac outlet, the reading between A and B and between B and C should be approximately 117 volts. The reading between A and C should be zero. For a 240-volt ac outlet, the reading between A and B should be approximately 234 volts, and the reading between A and C or B and C should be 117 volts.

of a battery-operated transistor radio. You can measure the voltage of the battery or of points in the circuit by putting the common (ground, or black) probe of your meter on the negative side of the battery, by moving the red (hot) probe to the positive terminal of the battery or to other random circuits in the radio. Resistance may be read by disconnecting the battery (never read resistance while voltage is being applied), and connecting the two probes to any two different points in the radio circuit board.

In order to determine the amount of current used when the radio is operational, you must disconnect one terminal of the battery from the radio, while leaving the other battery terminal properly connected. Say you have disconnected the ground end of the battery, while leaving the positive terminal connected. Now be sure the meter selector is in its highest dc current position. Then connect the black (negative or common) probe to the negative terminal of the battery, and the red (hot) probe to the point in the radio where the negative battery terminal normally connects. The voltage from the battery will force a current to flow through the now completed circuit — from negative terminal of the battery through the on switch, then on through the radio, through the meter, and back to the hot or positive terminal of the battery. In this process, the current flow will cause the meter to read, and you will be able to determine the amount of current in amps, milliamps (thousandths of an amp) or microamps (millionths of an amp) that is being used to make the radio operate.

Most VOMs are capable of reading current only in the milliamp range. This is fine for measuring the amount of current consumed by a transistor radio. It's not enough to test for current in a "Fry Daddy." Another kind of meter, and an expensive one, is needed for that.

Fortunately, current measurements are seldom necessary in appliance repair. If you can quickly and accurately make resistance and voltage measurements, you will be competent to do most of the diagnostic work involved in basic small appliance troubleshooting.

Chapter 2
Safety

It's tempting to get out the tools and tear into that malfunctioning appliance. But even if you're an old hand at handling electricity and machines that use electricity, take a few moments to review the basics given in this chapter. (If you *are* an old hand, you'll also realize just how important this review is.)

If all appliances were operated by battery, there would be no real danger. The dc voltage from a battery isn't nearly as dangerous as is the ac voltage used by your appliances.

Some years ago the United States Navy responded to the potential dangers of ac voltage by conducting a very thorough study of the subject, particularly the effects of current on the human body. The results of the study showed that it takes surprisingly little current for ac voltage to cause harm. Just 1 milliamp (1/1000th of an amp) is enough to be felt. 10 milliamps (1/100th of an amp) is sufficient to cause muscle spasms and paralysis. At this point, the person being shocked will be unable to let go of the source of the shock, and is likely to actually grip it more tightly. If the current is increased to 100 milliamps (1/10th of an amp) and continues for more than just one second, the usual result is death. The most important muscle in your body, your heart, will become paralyzed. You can't live long under those circumstances.

This isn't meant to scare you—although it should. The purpose is merely to let you know that you are definitely vulnerable. A flow of electricity of just 1/10th of an amp can be fatal, and you're dealing with a much greater flow than just a tenth of an amp.

A standard wall outlet in your home will be protected by a circuit breaker or fuse that can handle 15 amps. Other outlets might be going through a 20 amp or larger breaker or fuse. Large appliances will normally go through a circuit breaker or fuse capable of handling between 30 and 100 amps, depending on the appliance.

Even that relatively small outlet, then, can easily carry 150 times the current needed to be fatal. And that is for an almost indefinite period of time. Before that breaker or fuse lets go, or the wires in the walls melt, the current flow for a few seconds can be almost limitless — thousands of times more than your body can tolerate.

There are other dangers as well, such as heat and the risk of injuring yourself physically through carelessness. As has been mentioned before, you *can* successfully handle most repairs yourself. You can also handle them in perfect safety if you just pay attention to a few basic rules.

The money you've saved in doing your own repairs will be of little benefit (except to your heirs) if a careless accident while working on a device costs your life! For this reason, you *must* learn and consistently observe a few sound rules of technical safety before attempting to work on any appliance.

SAFETY RULES

Safety is really nothing more than a matter of common sense care. The rules of safety have been designed to prevent accidental injury.

General practices in professional machine shops and those followed by electrical workers are good ones for the home do-it-yourselfer. The safety tips listed in this chapter should be reviewed frequently and observed at all times while working on small appliances.

There are two basic reasons for observing the rules of safety. The first and most important is to protect your own safety. If you are careless and injure yourself, then you shouldn't be a do-it-yourselfer. Leave the job to someone else. You can replace the $50 you spend. You can even replace the entire appliance if need be. You *can't* replace an eye or a life.

Of secondary importance, but still important, is the safety of the appliance on which you are working. It doesn't do much good to save $30 or so in making the repair, only to waste $100 from damage you've done.

Quite often the two are interlinked. What protects you will often protect the appliance. This makes it doubly important to learn the rules of safety, and practice them until they become the natural way of working.

The following is a list of the most important factors for personal safety. Read them carefully. Better yet, read this section carefully and then post a copy of the list on page 32 in your work area and read that each time before you begin work.

1. Dress properly. Wear full-length pants covering your legs. No shorts, skirts, or clothes that have holes in them. A canvas lab apron can be used to cover your clothing if you wish to protect it, but it's best to work in old clothes that can be easily discarded if accidentally torn or stained. The complete covering from pants or a lab apron or overalls is to protect you from burns or shock while soldering or using electric equipment.

It is safer to wear rubber-soled shoes or to stand on a rubber pad when working with electricity. *Never* go barefoot, stand on concrete, or on the ground outside while working with anything electrical. Also, never work with electricity in an area where water has spilled on the floor or while standing on damp earth.

Protect your eyes. It's a good idea to wear shatter-proof glasses or safety goggles when soldering or working with motors that spin at high speed. Specks of grit or metal or hot solder flicked into an eye can permanently blind a worker.

2. Remove all jewelry. Metal jewelry is an excellent conductor of electricity. Keeping it on while working with electricity can lead to a harmful or fatal shock. Also, jewelry and loose belts or a tie can quickly become caught in moving machinery or gears when you are involved in repair work. This can lead to the loss of limbs, or loss of life if the equipment is big and powerful enough.

3. Follow the one-hand rule. Just as current flows from one side of a wire to the other, it can flow through your body. If both your hands come into contact with the circuit, the current flows through the body — across the chest and heart regions — which is very dangerous. Always keep one hand in your pocket or behind your back when measuring current or voltage. Some workers will use a metallic band around the wrist of the working hand, with a wire constantly attached from that band to electrical ground. Then, if a finger, knuckle, or elbow accidentally touches a hot line, the current runs through the shortest path to ground — through the hand or arm, to the wristband and its connecting wire to ground.

4. No food or drink. Do not work while consuming food or liquids. Spilled liquid becomes a short circuit path that can damage equipment, and spilled liquid or food can become a corrosive agent that can damage the appliance sometime in the future. Having food or drink nearby while you are working is also a distraction. It's more difficult to concentrate on probing that wire if you have a soft drink at hand. If you're hungry or thirsty, it's time to take a break. Go in the house, and have your snack while enjoying some television (or while reading the applicable sections in this book).

Never, *never* consume alcoholic beverages while working around electricity or power tools! Alcohol is a depressant. Your reactions and coordination are both lessened. Even a single beer can reduce your abilities enough to create a danger, and often you won't realize it until it's too late.

5. Keep the work area clean. The floors and desk or table top in your work area must be kept clean and free of any litter that might cause someone to slip, trip, or stumble. Some litter, if metallic, could even cause shock or a fire-generating spark if it hits "live" electrical circuits. A buildup of dust and dirt can make the job more difficult. It can also get into the appliance and create a number of future problems.

6. Have the proper work attitude. Don't clown around or engage in horseplay while using tools or working with power-driven devices. Many painful injuries are caused by the careless and thoughtless antics of a would-be comedian. Do not talk to or distract anyone when they are working with electrical devices or with potentially dangerous tools. And don't let anyone do it to you. If your long lost brother walks in while you're pulling out a heating element, stop what you're doing and give him your full attention. Or ignore him and give the appliance your full attention. You can't safely have it both ways.

7. Use the correct tool. Each tool has a particular function. Don't attempt to hammer with the handle of a screwdriver or use the point of a knife to turn a screw. Slips in attempting to use the wrong tool for a job are a major cause of cuts and bruises in workshops. And do not use *any* of your tools if they are not in proper working condition. Additionally, it is important to use the right *size* tool for the job at hand. It might be tempting to remove a screw with the closest screwdriver, regardless of blade size. Don't.

8. Cut away from the body. When using any type of cutting tool, make sure to slice or cut *away* from your body. If you slip, about the worst that will happen is that you'll damage the thing being cut. But at least you won't find yourself sitting there at the work bench trying to figure out how to get a knife out of your leg.

9. Take care of injuries immediately! Take competent care of any injury that does occur at once. Even the slightest cut or burn can develop serious complications if not properly treated in time.

No workshop, and certainly no home, should be without a complete first aid kit. Everyone in your home should also know how to use the kit. A first aid kit is all but useless if no one knows how to use it. (Worse yet are those homes that have a first aid kit and no one even knows where it is!)

Every area in the country has classes available in first aid, up to and including CPR courses. These classes are either free or very inexpensive. An accidental electric shock that stops your heart need not be fatal — *if* someone in your home knows how to handle the emergency.

10. Unplug! Before you begin working on an appliance, be sure that the plug has been removed from the outlet. It's not good enough just to shut off the switch. There will be times when you'll have to have power applied to carry out a test. That's fine. In this case, your conscious and deliberate action will be to plug in the appliance. After the test is done, unplugging it again should be such an automatic reaction that you hardly realize that you've done it.

Many a finger has been lost by repair persons who took apart a disconnected fan to effect a repair, fixed the fan motor, plugged it in to test it, turned it off, then forgot to unplug the appliance before putting it back together. Accidentally hitting the power switch, the blades can swiftly become a human meat slicer!

Unplugging also applies to power tools. Never leave a soldering iron or a piece of electrical equipment under repair while it is still turned on or plugged in to a wall outlet. Don't leave the work area for any reason until you have double checked to make sure that all heat sources have been disconnected and that all potentially dangerous items have been put away in their proper place.

Young children are adept at finding an opportunity to play with dangerous power tools or around electrical appliances that may contain short-circuits or mechanical malfunctions. If you can't fix it right away, put it away!

11. When in doubt — DON'T. The line voltage that powers the appliance is dangerous. Even more dangerous is overconfidence. If a situation comes up and you don't really know how to handle it, stop! It's time to think things over. And if you can't figure it out, call in a professional.

Safety Rules

1. *Dress properly*	6. *Keep the work area clean*
2. *Remove jewelry*	7. *Use the correct tool*
3. *Use the one-hand rule*	8. *Cut away from the body*
4. *No food or drink*	9. *Take care of injuries*
5. *Have the proper work attitude*	10. *Unplug!*
	11. *When in doubt — DON'T*

DANGER SPOTS

Most of the dangerous spots in an appliance are obvious. By now you realize that anywhere that ac line voltage is present is a spot to be avoided. Avoiding anything hot is another obvious danger. Then there are the various mechanical dangers, such as sharp metal edges, screws, and fasteners, and various parts that may suddenly snap or shatter. However, not all danger spots are quite as obvious.

If one of the ac lines has been damaged in some way, it could be touching the metal chassis. This can also happen if you accidentally (or purposely!) cause a short between the incoming line voltage and the chassis. In essence, this could make the entire appliance a dangerous ac source. You can't see if voltage is present. The simple solution is to unplug the appliance before you work on it.

Heating elements are used in many appliances. Even those that don't actually have a heating element can have parts that get hot during operation. Many people make the same mistake with heat as they do with electricity. They expect to be able to see it. Although it is sometimes visible, such as when a heating coil glows, it isn't always. If you can safely do so, and without actually touching anything, move your hand over the top of any suspected hot components. Heat rises. If heat is present you should be able to feel it.

There is another danger that many people never think about, until they've experienced that danger first hand. This danger involves capacitors, or "cans" (so-called because that's just what they look like). Capaci-

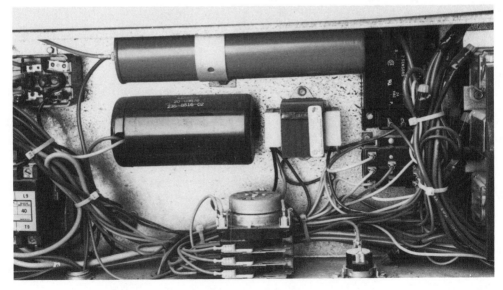

FIG. 2-1   A large capacitor can hold a dangerous charge.

tors are electronic devices which can be used in several ways. One use is to separate or "filter" different types of current — ac and dc. A capacitor blocks dc, while allowing ac to pass easily. If a particular appliance is prone to RF (radio frequency) interference, capacitors can be used to pull that RF signal away from the appliance and to a ground. These capacitors are usually very small and present no danger.

The capacitor can also serve as a sort of temporary battery, storing a charge until it is needed. If the applicance has some sort of electronic circuit board with transistors or IC chips, then there will be a power supply for the device, which converts wall outlet ac to the dc needed to operate these solid-state devices. Capacitors are a major component of power supplies and are used to smooth the voltage fluctuation. The partially converted incoming voltage flows into the capacitor where it is stored and then released in a steadier flow. This works fine as long as the power is flowing; and if everything is working properly, the capacitor will drain itself of charge through a resistor after the power has been shut off.

Another use of a capacitor also takes advantage of the capacitor's ability to hold a charge until it is needed. Electric motors and compressors draw large amounts of current when they first start up. This extra drain presents two problems. First, the fuse or circuit breaker could blow. Second, since a relatively insufficient amount of current is flowing at startup, the motor or compressor can't do its job as well and will wear itself out more quickly. A startup capacitor solves these problems by providing a boost.

However, any capacitor can hold a charge for a long time, long after power has been turned off or after a power cord has been disconnected from the wall. Make it a point to short out large value capacitors with an insulated screwdriver (after the power cord is disconnected), before attempting to work on the appliance. To short out the capacitor, place the metal shaft of the screwdriver against one terminal of the capacitor and simultaneously touch the point of the screwdriver to ground, and then to repeat the process at the other terminal. Any charge left will short circuit harmlessly to ground, instead of later jumping through your hand to your chest.

There is another danger well worth mentioning here — a fire hazard in *every* home which few are aware of, and which few take preventive maintenance steps to avoid. Household dust, especially when it becomes moist, can be a very effective conductor of electricity.

All wall outlets, light switches, and overhead and wall light fixtures are mounted in small metal or plastic boxes called electrical cases. These boxes are screwed or nailed to the studs or framework of the house, and then the outlet, switch, or light fixture is mounted over the case. The exposed (stripped of insulation) wire is wrapped around screw taps or solder connections, providing an electrical path to the outlet. Over periods of months and years, the outlet boxes become filled with dust. Some dust particles are *flammable*, as well as conductive. When an appliance is plugged in or out or a switch turned on or off, there can be a slight electrical spark when current first reaches the contact. This is caused when the contact is slightly corroded or otherwise not as solid as it should be. The spark that results is the *number one cause* of household fires in the United States. The most frequent occurence is when an iron or a space heater, drawing heavy current, is pulled from a wall outlet or plugged into an outlet while the appliance is *on*. A spark naturally occurs when electricity jumps to the circuit being completed or opened by insertion or withdrawal of the plug.

Never plug an appliance in or remove a wall plug unless the power switch for that appliance has been turned to the off position. And it's a good habit once a year to remove all plastic wall plates covering outlets, switches and fixtures. Carefully use a plastic (insulated) extension on a vacuum cleaner to dust out the outlet boxes. If the vacuum attachment won't reach into the small areas, blow hard to get loose dust out, or use a small air compressor to blow out the dirt and grime.

APPLIANCE SAFETY

After you've made sure that *you* are safe, you can consider the safety of the appliance. All that is really needed is common sense.

If the appliance you are lifting is heavy, protect yourself and the appliance by getting help. Even then, be sure that you and your partner are using the proper lifting technique. Lift with the legs and not with the back. If you're bent over, you're not doing it right. The correct motion is an almost straight up-down motion, with the force being supplied by your legs.

If you feel the appliance slipping from your grip, don't be too proud to admit to it (immediately!) and set the appliance down again for a better grip, or a rest, or to simply abandon the attempt until even more help can be secured.

One of the greatest difficulties that both the novice and the old pro run into is that of removing or replacing the fasteners. Screw heads strip, making it difficult or impossible to move them. Worse yet, a screw or bolt can break.

In replacing those fasteners, many people have difficulty in getting the screw into the hole. So it goes in at a strange angle and is forced to cut new threads. This is *not* a valid function for a screw. Take your time, use the right tool for the job, and do it right. The day could come when you'll want to remove that screw again. (You also don't want to inadvertently have the screw cut into a wire.)

Some appliances have hidden fasteners. Others use plastic clips to hold things in place. With both, it's easy to damage the appliance— sometimes beyond repair—if you're not careful.

If the part you're trying to take off doesn't seem to want to move, don't force it. Look carefully for something else that's holding it. (See Chapter 4 for more information on disassembly and assembly.)

FIG. 2-2 Look for hidden screws and fasteners.

Chapter 3
Fundamentals of Electricity

To better protect yourself and the appliance, and to make the job easier, you should have at least a basic understanding of what electricity is, and how it does what it does. This is not a difficult or complicated subject. Most of it is merely a matter of understanding the terminology and learning the definitions of the words used. (To help with this, a glossary is provided at the end of this book.)

It's important to know the terminology. Without it you might be able to fix the appliance well enough, but you won't be able to communicate very well should you need parts or the advice of a professional. ("I need one of those round things that screws into that square-kinda thing over in the lower back corner by that long skinny thing.")

Knowing a little about how electricity makes things work can also help you in other ways. First and most important, when you understand how electricity does the job, you are less likely to be injured by it. Second, your understanding can also make it much easier to diagnose and handle malfunctions.

The main thing to keep in mind is that electricity always requires a complete path to work. (It's surprising how often even seasoned professionals forget this simple rule.) This is true for both ac and dc.

Think of a battery, such as the one you use in a flashlight. It has two sides. One is flat (the negative side), and the other has a small bump sticking up from it (positive). If you connect just one side of the battery to a device, nothing is going to happen.

The electricity often passes through a number of wires and components. If a break happens anywhere in that path, things can come to a grinding halt. At the very least, the way things operate in the appliance is going to change.

If everything stops working, nearly every time you'll find the problem in just one of two areas. Either the power isn't getting to the device in the first place, or there is a break in the pathway. This break could be something simple like a broken wire or blown fuse. It could also be from a faulty component that won't allow current to pass.

FROM THE SERVICE ENTRANCE IN

Power is carried along the heavy wires of the power company, through the large transformer that changes the primary voltage into the voltage needed in the home, and then through the meter and into the service entrance of your home. This is the main breaker or fuse box. There may also be one or more secondary boxes.

Everything up to the main breaker box is in the realm of the power company. Under no circumstances touch anything there. If there is a problem on that side of your service entrance, call the power company immediately and let them handle the rest.

From the service entrance and into the house is your responsibility. Local ordinances might make it illegal for you to work on the wiring, but

FIG. 3-1   The service entrance. Note the circuit breakers.

Chilton's Guide to Small Appliance Repair/Maintenance
FUNDAMENTALS OF ELECTRICITY

that just means that you must pay the cost of having a licensed electrician come out.

Total service amperage for most home wiring circuits is 100 to 200 amps, depending upon local housing codes. Most of the fuses or circuit breakers are 15 to 20 amp, and they control those circuits involving other household wiring such as to lights and wall outlets. Heavier 20- to 40-amp fuses or breakers are used for the circuits feeding 240-volt devices such as stoves, air conditioners, clothes dryers, etc.

The service entrance is always supplied with a main switch or main breaker, which can disconnect electrical service to the entire house. It can also be used to cut off the power supply from the utility company to the breaker box so that circuits can be rewired or new circuits added.

BRANCH CIRCUITS

From the main box, the incoming power is divided (distributed) across a number of individual circuits. As mentioned above, these circuits are controlled, and protected, by fuses or circuit breakers. Usually wires connect together a number of outlets or lights. You might have both lights and outlets on a single circuit controlled by the same breaker.

It's important that you label these branch circuits at the service entrance. If they're not already labeled, it's going to take some time and effort on your part to figure out what goes where.

Do this by a process of elimination. One way is to flip off a breaker and find out what outlets or fixtures are no longer powered. The other way is to shut all the breakers off and then flip them back on again one at a time to find out what *is* powered. This will also involve the use of a VOM to test wall outlets.

Expect to spend the better part of a day at this, and lots of running in and out. A piece of masking tape stuck on the cover plate of each outlet after it shows power both eliminates that outlet from future testing and gives you a place to write a number or other code. As the job of tracking everything down continues, it then becomes easier and easier since there are fewer places to test.

If there are 30 outlets in your home, the first test through might eliminate 5 of them. Mark those and the next run through the house with VOM in hand will give you only 25 to test. And so on until all have been marked.

Leave the lights switched on throughout the testings until they have been eliminated. This way you'll know which breakers operate which lights and which switches.

You'll probably find that there is a logical order to the way things are wired in your home. An outlet in the kitchen is unlikely to be powered by the same breaker as one in the garage, for example. A room with multiple outlets and lights (which will be almost all rooms in your home) will probably not run all from the same breaker.

FUSES AND CIRCUIT BREAKERS

Despite all precautions, it's possible for a short circuit or other malfunction to occur in an appliance. If this causes a heavy flow of current, the wires in the wall can heat up and eventually melt. A fire could start. After this you won't just lose the appliance—you could lose your home, or worse.

Fuses and circuit breakers are safety devices to prevent overloads and to keep short circuits from causing wires to overheat to the point where they become fire hazards. Fuses are rated according to the amount

FIG. 3-2 A fuse and a circuit breaker.

of current (amps) they can carry. When a circuit attempts to draw more current than the rated fuse value, it will burn out.

A circuit breaker is much like an automatic switch. It is a modern replacement for old style fuses. Like a fuse, it will automatically turn off whenever a circuit overload or short circuit tries to draw more current than an electrical device is supposed to use. Unlike a fuse, the circuit breaker needs only to be reset to return electricity to the circuit.

CHECKING FOR OVERLOAD

When too many appliances or lights are placed on a single electrical circuit the result is an overload. For example, a 15-amp circuit may be used to operate wall outlets in the living room and a hallway of a home. If lights in use are drawing a total of 7 amps, a TV set is on drawing one-half an amp, and someone then plugs in an electric iron that draws 10 amps while heating—that's a total of 17½ amps on a 15-amp circuit. The circuit will be overloaded. Some of the lights will have to be turned off in order to use the iron there, or the iron will have to be moved to another room or plugged into an outlet from another less-used circuit in order to prevent the overload.

A blown or burned out fuse indicates that you have placed an overload on that circuit, or that some part of the wiring within the house or within an appliance plugged into one of the wall outlets has developed a short circuit. Visual evidence is readily available for a circuit where the safety device has tripped. If a plug-type fuse has blown, and you can see the open fuse strip through the small mica window, it is generally an indication of a circuit overload. If the fuse burned out due to a great deal of heat, blackening the window, it usually means a short circuit has occurred.

Cartridge-type fuses must be checked for continuity on the resistance (ohms) scale of your VOM. Remove the fuse from its holder and probe it with your meter. If the meter shows no resistance (continuity) from one end to the other, it is still good. If it shows infinite resistance (an open circuit), the cartridge fuse is blown and must be replaced.

With blown screw-in plug-type fuses, a 25-watt light bulb may be used to test a circuit. When the fuse blows, turn off the main or master power switch in the service box. Unscrew the blown fuse and replace it with the 25-watt bulb. Put the main switch back into the on position. If the bulb burns quite dimly, there is an overload and some appliance or light must be disconnected before the circuit can again work properly. If the bulb burns brightly, there is a short circuit. Go back into the house and disconnect lights and appliances one at a time, going back to the fuse box each time to check the condition of the bulb. When unplugging an appliance or lamp causes the bulb to go out, you have found the unit containing the short circuit. Leave that appliance disconnected, turn off the main power, unscrew the light bulb and replace it with a new fuse. Turn the main switch back on, and take the faulty appliance to your workbench for a checkup.

With circuit breakers, there is either a "red flag" (a red piece of metal that shows through a window when a breaker is tripped) or the breaker switch will be moved over towards the "off" position. (With some breakers this movement is small.) To reset, just move the toggle switch all the way to the off position, then back to the on position.

No matter what type of protection is used in the wiring circuits of your home, try to determine what caused the fuse or breaker to go before replacing or resetting. Disconnect some of the lights or appliances on that circuit, in case it was an overload. If the circuit then works fine when reset or with a replacement fuse, you might—one at a time—turn on each of the lights or appliances that were active when the first overload occurred. This will help you determine if some of the appliances might have to be moved to another home circuit.

If the new fuse continues to blow or the breaker keeps tripping even after all appliances, lamps, and ceiling lights have been turned off, the problem is probably a short circuit in the wiring itself. You will either have to get professional help from an electrician, or use your VOM to try tracing down that short circuit so you can fix it yourself.

If the breaker (or fuse) holds with all lights off and all appliances disconnected, then—one at a time—turn on the light fixtures. If the safety device blows when you turn on one of the lights, the short is within that lamp or its on/off circuit or its power cord or plug—or within the bulb itself. To effect a repair, see the Lamp servicing tips in Section Two of this book.

If the breaker holds for all the lights, then carefully plug in and then turn on each of the appliances on the circuit. This way, you can discover which appliance was at fault and can begin repair work on that particular appliance.

In some instances, everything might work fine after the breaker has been reset or the fuse replaced. The short or overload which caused the trip could have been a temporary malfunction or a power surge from the utility company. This is particularly possible in periods of electrical storms, or when excessive electrical use by all consumers in a service area is causing lights to dim or flicker.

A note of extreme caution. *Never* replace a fuse with a new one of a higher amperage rating, or by using a penny or piece of metal as a temporary replacement. This is a highly dangerous fire hazard. Always use replacements of the exact same rating. Do not attempt to change fuses or work on a breaker box in the dark, or when standing on wet ground. Keep a flashlight handy on top of or beside the service entrance box, and stand on a dry, rubber mat if it has been raining or the ground is damp. It is a wise precaution to keep one hand in your pocket when working with fuses or breaker boxes. Throw the main power switch into the off position before working with individual circuits, and have someone else around to assist with first aid or in turning off the power in case of an accident.

HOME WIRING

In this section on electrical wiring, you will find a basic review of wiring circuits, a description of many of the terms used, working hints when handling electrical circuits and wiring, and a review of some specific problems and their cure. Working with electrical appliances is a lot easier, and a lot cleaner, than most mechanical work. But it is more dangerous if you become careless or rushed. In a few communities in this country, only professional electricians are authorized to do *any* repair or installation of home wiring circuits. In most areas, though, it's permissible for the home handyman to do the electrical work and then have it inspected by local authorities.

If any rewiring is to be done, whether the local code calls for an inspection or not, the work must conform to standard wiring codes and practices. Some of these rules are listed in this chapter. You should ask your local building authority (inspector) for a copy of the local code before undertaking any chores involving home wiring. Another good source of information is the National Electrical Code. It costs $15.00, and a copy can be obtained by writing to the National Fire Protection Association, Inc., Batterymarch Park, Quincy, MA 02269.

Even if local regulations make it illegal for you to tackle the wiring in your home, the home handyperson can, of course, work on any home appliance without calling in inspectors or checking with outside authorities. Often, however, the addition of a new appliance to a home necessitates a change in the wiring circuits from the fuse or breaker box or the installation of more wall outlets.

Occasionally, it is necessary to provide a 240-volt outlet (instead of the normal 120-volt outlet) for a window air conditioner or some other high-wattage (high power consumption) appliance. In these instances, you can either hire a journeyman electrician or electrical contractor or you can do it yourself following the guidelines of this book. If you do it

Common Wire Types

TYPE	USE/COMMENT
AC	Commonly called BX; metal-armored cable
HPD	Asbestos covered wire; heat resistant
NM	Multi-wire cable
NMC	Same as NM, but moisture- and corrosion-resistant
S	Flexible cord, with stranded wire
SO	Same as S, but oil-resistant
SP	Flexible cord, rubber-coated
SPT	Flexible cord, such as used for small appliances
T	Solid wire with plastic insulation
TW	Same as T but moisture-resistant
THW	Same as T but moisture- and heat-resistant

yourself, get an inspection by the city building authorities after you have completed the work.

Again, check the local codes. In most instances you must obtain a building or an electrical work permit before starting the job.

TYPES OF WIRE

The type of wire used will depend upon your local electrical code and on where and how the wire will be used. For home wiring, the most commonly used is Type T. This is a single solid wire coated with plastic insulation. If the wire is to be in a wet area, then Type TM is used; if heat is present, Type TMH is used.

An easy way to run multiple wires is to use a cable assembly. Inside the cable are several wires, each with its own insulation. The designation on these cables is usually given as a number, such as 12-3 (3 wires of 12 gauge) or 14-2 (2 wires of 14 gauge).

Some electricians use metal-armored cable, commonly called BX cable but more correctly called Type AC. BX cable cannot be used except in areas that are always dry. The Romex insulation is either rubber or a heavy plastic coating. For use outside or in areas likely to become wet, building codes require that Romex be installed through a sealed conduit or pipe (either galvanized metal or PVC plastic piping). In any household work, it is best to use the same type of cable already in use for similar circuits in your home. Special grounding circuits are necessary when connecting Romex to BX, or BX to Romex circuits.

The size (gauge) of wire to be used depends upon the current it will be required to carry. The *smaller* the gauge number, the *larger* the wire. Eight-gauge wire is more than ⅛ of an inch in diameter, while 14-gauge wire is less than 1/32 of an inch thick. The electrical cord supplying voltage to most appliances is either 12, 14 or 16 gauge. Irons, microwave ovens, toasters, broilers and units using more power (those containing heating elements) draw more current and need a larger (lower gauge number) power cord than electric razors, lamps, or radios that draw relatively small amounts of current.

Most home circuits work off a 15- or 20-amp fuse or circuit breaker. The wire for such a circuit must be of at least 12 gauge. If you are installing a new wall outlet to be used as a power source for a microwave oven or for a small room heater or air conditioner, it is best to use 10- or 12-gauge Romex or larger, since the appliance may use almost all of the 15 amps from the house fuse or circuit breaker, especially when first turned on. It is best to isolate these circuits to only one or two outlets, since an overload will result if you try to operate too many other appliances on the same circuit that feeds a high-wattage device.

For installation of wiring for a heavy-duty air conditioner or heater, consult local codes and experts. It may be necessary to install a circuit of a higher amperage for correct operation—for instance, a circuit breaker or fused circuit of 20 or 30 amps instead of the normal standard 15 amps. For this, wiring codes definitely specify 10-gauge or larger Romex feeding from the circuit box to the wall outlet. Use of too small a wire will lead to overheating of the Romex, and a potential fire hazard. This is what led the National Fire Underwriters Board to develop electrical codes. At times the specifics required of an installation might seem petty, but don't ignore them.

Electrical codes specify not just the size of the wire but also the color of the insulation. This is to help protect the people working with those wires. (Imagine opening a box and seeing a collection of colored wires, with no idea of what any of them are carrying.)

| ∅ | 2 | 4 | 6 | 8 | 10 | 12 | 14 | 16 | 18 | 20 |

FIG. 3-3 Wire gauges.

As has been mentioned, a complete path is needed for electricity to work. This means that one wire brings the current to the outlet, and another takes it back. The one bringing in the current is the "hot" wire; the one taking it back is called the return, the ground, or the neutral.

The grounded wire is always white (assuming that the installer followed code). This wire can never be switched, fused, run through a circuit breaker, or interrupted in any way. The only time you won't find a grounded wire is when the outlet is supplying current for something operating only at 240 volts.

The "hot" wire for ac voltage is usually black. By code it can be anything other than white or green in color (and certainly can never be a bare wire).

Most modern wiring circuits use different metallic colorings. The "hot" side is generally considered to be the copper or brass-colored contact. The white-colored contacts, which are sometimes coated with nickel or tin to become silvery, are used for the white wires, or the ground side of the circuit. (For a socket, the white wire goes to the screw shell, with the "hot" connected to the center contact of that socket.) A screw

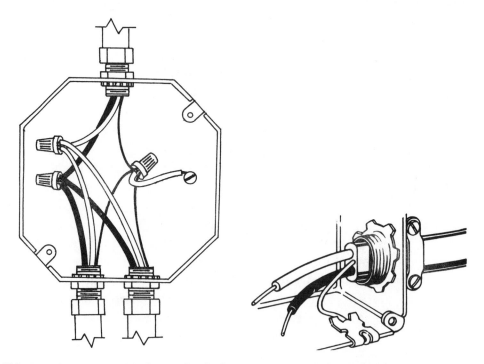

FIG. 3-4 How a junction box is hooked up. One wire is "hot"; the second is used as a neutral or return and is connected to the ground of the power company. The third wire is connected directly to ground to carry away voltage. Note that the "hot" wire is black. Reprinted by permission from *Chilton's Home Wiring and Lighting Guide* by L. Donald Meyers (Radnor, Pa.: Chilton, 1980).

contact that is green is the grounding wire, sometimes called the safety ground. The wire to it will be green, green with yellow stripes, or can be a bare wire.

It might sound confusing, but there is a distinct difference between the grounded wire and the grounding wire. The grounded wire does carry current, but it carries it back to the service entrance and into the power company's system. The grounding wire—to the green contact, and what is often a bare wire—doesn't carry any current at all during normal operation. Instead it serves as a safety and pulls the current away from the outlet or appliance and literally into the ground. Besides being used to pull the dangerous current away from an outlet, it is also directly connected to the metal chassis of an appliance. Then if something goes wrong, the metal body of the chassis or its parts (such as a motor) cannot hide a deadly charge because it has been discharged by the grounding wire.

All this assumes. of course, that the wiring has been done correctly. It should also give you a good idea as to why color coding, and paying attention to the codes in general, is so important. Open an electrical box for ac voltage, see a black wire, and you *know* that it's deadly unless you stop the power outside the house.

SWITCHES

Switches are lever, push-button, or revolving rheostat-type devices to make or break an electrical line; that is, they open or close a wiring cir-

FIG. 3-5 A typical wall switch.

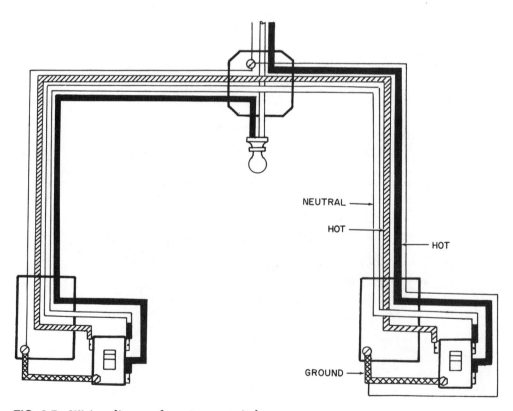

FIG. 3-6 Schematics of SPST and DPST switches.

cuit in order to turn an appliance on or off. With the circuit *closed*, the switch is on and the circuit is energized. An *open* circuit results when the switch is in the off position, and no current flows to the appliance.

Most wall switches in a home are of the single-pole, single-throw type (SPST). This means the switch is in use across only one line and either on or off. It is connected in series with the hot wire of a circuit (the black wire according to U.S. electrical wiring codes, although any color other than white or green may be used).

A double-pole, single-throw (DPST) switch has four contacts. It splits both the hot and the ground wires of a circuit. One side of the switch is in series with the white, or ground wire; the other side of the

NEUTRAL

HOT

HOT

GROUND

FIG. 3-7 Wiring diagram for a 3-way switch.

switch opens or closes the hot or black wire. The DPST opens or closes both wires when it is operated.

In cases where two switches are used to control a single circuit (an inside and an outside switch for a porch light or a garage door-opener), a three-way switch is necessary. This type of dual control requires a three-wire cable between the two switches and the device to be operated.

In addition to wire and switches, there are outlet sockets, outlet boxes, switch or junction boxes, and connectors to secure the Romex or BX cable to the boxes.

Switch boxes, rectangular in shape, are made of galvanized metal or plastic and are used as the housing for switches or wall outlet plugs.

Junction boxes are octagonal in shape and are used for joining (splicing) wires from different circuits, or for mounting fixtures such as a ceiling light.

The outlets are either one-, two-, or four-plug sockets, designed to provide electricity when an appliance is plugged in.

For new installations, you will also need solderless connectors, which are twist insulated caps that "screw" over the ends of two stripped wires that have been twisted together (spliced), electrical tape, and either straps or insulated staples to hold cable to wall and studs as it is being strung from breaker box to final location.

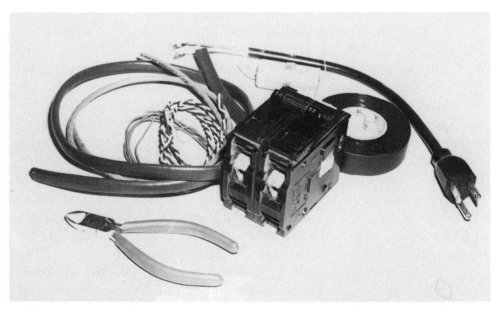

FIG. 3-8 Electrical supplies. From left, wire cutters, wires and cords, fuses and breakers, and electrical tape.

TESTING AND REPLACING ELECTRICAL OUTLETS

It's likely that sooner or later you will need to replace an outlet. Over the years, inserting and removing a plug from the outlet will wear it out. Beyond that, and despite everyone's efforts at quality (hah!), things do go bad. If no power is coming to the outlet, yet the fuse or breaker and the wires are good, then it's time to replace the outlet.

Testing an outlet is easy. You need nothing more than your VOM, set to read in the 120 vac range. Since you're working with ac, it doesn't matter which probe you use for which hole. Just make sure that you hold the probes *only* by the insulated handles.

First, put a probe into each of the two flat slots. The meter should show you a reading very near 117 volts. Next, put one probe into the larger of the two flat slots, and the other probe into the round hole beneath. There should be no reading at all. If you *do* get a reading, the outlet or wiring has either been hooked up wrong in the first place, or has shorted. Finally, touch one probe to that same round hole and move the other probe to the second flat slot. You should once again get a reading of 117 volts ac.

This test shows that the narrower slot is bringing the needed ac to the outlet, that the ground*ed* wire is carrying it back and completing the circuit, and that the ground*ing* (the rounded hole) is functioning as a safety.

It is recommended that you carry out the test several times, especially if you're not 100% comfortable in using the meter.

The 117-volt reading is an average. It will fluctuate. A reading of anything from 104 to 127 is considered normal. A difference of a few volts won't matter. However, you may get consistently high or low readings every time you test. This condition might require correction. Contact the power company.

Some appliances will not easily tolerate changes in voltage. For example, a microwave oven will not cook properly if the supply voltage is too high. If the wall outlet used to feed your microwave consistently shows an output voltage of 125 to 127 volts, you might try operating the oven with a short (six foot or less) 10- or 12-gauge extension cord. This is a violation of a standard warning not to operate heavy-duty appliances with extension cords, and it is a poor solution. (You should call the power company.) But when the voltage supply is too high, the cord can provide the necessary voltage drop to bring the supply to the appliance down to the 117 or 120 volts needed to kick the microwave klystron into operation.

Motors and heaters will work inefficiently if the supply voltage is too low, and a low voltage will shorten the working life of these appli-

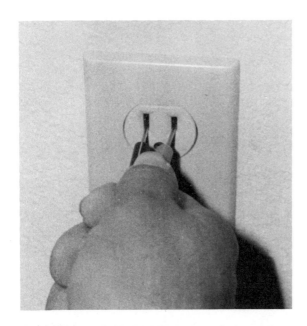

FIG. 3-9 To test a wall outlet, start by probing the two flat slots. The reading should be 117 volts.

FIG. 3-10 Next put one probe into the round hole and the other into each of the two flat slots. The reading between the round hole and small slot should be 117; between the round hole and the large slot, the reading should be zero.

FIG. 3-11 Changing an outlet.

ances. If the supply from the outlet to be used is consistently below 110 volts, then select another circuit for that appliance. It may be necessary to operate that particular appliance from a single (isolated) circuit, that is a dedicated circuit with no other electrical device plugged into it.

Replacing the outlet is easy. All you need is a screwdriver, although a knife or wire stripper might be needed if the end of the wire has been sparking and is charred.

First the most important step. SHUT OFF THE POWER AT THE FUSE OR BREAKER! If you're not sure which one it is (all fuses and breakers should be labeled at the service entrance), shut down all power coming into the house.

Remove the two screws that hold the cover plate and then lift that plate out of the way. Inside you'll see the outlet, the wires going to it, and two more screws that hold the outlet to the box. Remove these two screws.

You can now pull the outlet from the box. There is bound to be some stiffness to the wires, but don't yank on the outlet. A firm pull will get it out far enough.

Check once more that all power has been shut down. Although using a meter is fine for this, don't trust it. Go to the service entrance and double check that you have the power cut.

The wires should be color coded, as specified above. If you're uncertain, simply get out some masking tape and label the wires. If necessary, clip off the ends (as little as possible) and strip back insulation to bare clean, shiny wire.

All that's left now is to reverse the procedure. Put the wires in place on the new outlet, tighten the terminal screws completely so that no bare wire shows (which will mean bending a loop in the direction of the turn of the screw if you've had to clip the wire), carefully push the outlet into the box, tighten those two screws, put the cover plate on, and tighten those two screws.

Before you consider the job done, get out the meter and test the new outlet completely, both for power coming in and for the proper safety grounding.

SOLDERING

To do soldering correctly takes practice. And the place for that practice is *not* inside the appliance. If you've never worked with a soldering iron or gun before, practice first on some scrap wire. Learn how to do it properly before you attempt it where it will count.

Before you begin the job, be sure that the wires to be soldered are perfectly clean. If they're not, sooner or later you're going to have trouble. Clip off enough of the end of the wire so that the wire is clean and shiny.

When making a splice or tap, be sure that the joint is physically strong. Solder is meant to create a joint that will conduct easily and that will resist corrosion. It is *not* meant to weld or glue the wires together.

Hold the soldering tool beneath the joint. Once the joint is properly heated, it will melt the solder. Resist the temptation to melt the solder

Soldering Tips

1. *All wires must be perfectly clean.*
2. *Make the joint physically secure before you solder.*
3. *Apply heat to the joint, not to the solder. Let the joint melt the solder.*
4. *Use the solder sparingly.*
5. *Test the joint for continuity.*

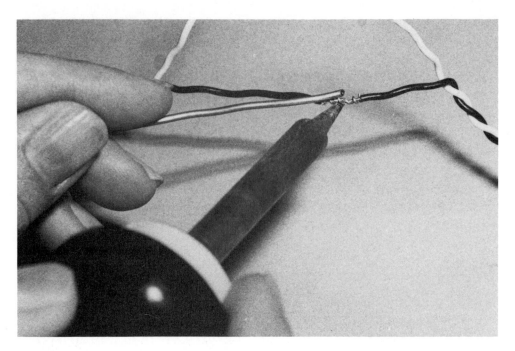

FIG. 3-12 Learn how to solder correctly.

with the soldering tool. If you do this you'll end up with what is called a cold joint. This means that the solder has not properly adhered to the wire, and you'll eventually have trouble with that joint.

If you look at the soldering work of a pro, you'll see that very little solder is used. There will be just enough there to "tin" the wires. You should never see blobs or drips of solder. If you've followed rule #3, the wires will melt the solder and it will flow evenly over the wires, turning them a shiny silver color. As soon as this happens, the job is done. Don't keep adding solder.

To make the job easier and more professional, take a few extra minutes and pre-tin the wires to be joined. This gives the wires a light coating of solder and will help to create a more secure joint.

REPAIRING WIRING AND CONNECTIONS

Not long ago we moved into a new home. It was then we discovered that our kitten had at one time or another decided that the lamp cord hidden beneath the couch was a wonderful chew-toy. Somehow the kitten survived and became a cat. But it meant that the lamp had to be rewired.

A number of things can happen that necessitate replacing a wire, cable, or power cord. Tugging at a sharp bend in the cord may have caused it to partially or completely break at the point of the bend. Rubbing action could have worn away or frayed the insulation, leaving wire exposed. In some instances, the exposed wiring could have led to a short circuit, causing the electricity to travel instantly from source back to ground. The jumping over of electricity from one point in a wire to another leads to an instantaneous demand for high current from the power source (a wall outlet), and can easily generate enough heat to burn a wire in half.

Another potential problem is that someone may have stepped on the plug and damaged it, or that one lead of the two-wire cord may have been pulled loose either at the plug or as the cord enters the appliance.

You, too, might have a kitten chew through a wire, forcing you to replace it (and possibly the cat).

When you need to replace a faulty power cord for an appliance, here are the recommended sizes:

1. For clothes dryers, stoves, and large air conditioners, use 3-wire (grounded) 6- or 8-gauge cable or power cords. Use two separate fuses or breakers (or a double breaker) for a 240-volt appliance, protecting each of the two black wires (the two hot sides of the line). The white or green wire is the neutral line, or the center tap of the 240-volt source, and it should go to the ground terminal of the plug and the ground terminal of the appliance.

2. For small kitchen appliances (toasters, broilers, mixers, blenders, etc.) use 10- or 12-gauge wire or cable. It is best to operate frequently used kitchen appliances off of a 20-amp circuit in order to prevent overloads.

3. For lamps, radios, TVs, and movable appliances such as the vacuum cleaner, replacement cords should be 2-wire 12- or 14-gauge. Small electric heaters need to be serviced by a 2-wire 12-gauge cable, on a 20-ampere circuit.

The size of wire used for replacement cords is not all that important unless that appliance is drawing a lot of current. The easiest way is simply to take the old cord (or a piece of it) along and match it for size with the new wire or cable that you buy.

Many appliances and power tools come from the manufacturer with extremely short cords. The idea is that the owner is less likely to slice through the cord, or to knock over the appliance, if the cord is short. However, in some cases the cord is so short that you have to use an extension cord just to be able to use the appliance or tool at all. Or you have to replace that supplied cord with a longer one. (It is *never* wise to attempt to operate irons, space heaters, stoves, microwave ovens, automatic irons (mangles), or air conditioners with an extension cord. The cord may overheat—a fire hazard—and a resultant low supply voltage may shorten the working life of the appliance.)

When replacing a cord with a longer one, or when using or repairing an extension cord, you must be careful not to go too long a distance. Voltage is lost if a wire is of too small a gauge over too long a run. If the wire fails to deliver the needed voltage to an appliance or power tool, not only will the device fail to work at top performance, but there is a danger that any electric motor and certain other components will burn out quickly.

There are two "most important" considerations involved in the repair and installation of wiring. First, the ends of the wires involved have to be bright and clean before they are connected. Copper wire should be shiny and bright. If it is dull or blackish in appearance, either thoroughly clean the wire until is is bright and shiny, or trim back farther until you get shiny wire. If need be, replace the whole wire. Settling for less than a perfectly clean wire is dangerous.

Second, the final connection must be physically solid and insulated. This can be accomplished by using screw clamps, soldered connections or solderless caps, and by covering the connection with electrical tape. For splices and taps, learn the proper methods for securing the wires together. Do this before you complete the job by soldering the joint.

Never work on an electrical appliance while it is plugged into a power outlet. Disconnect the plug, and place the defective unit on a table or workbench. If the cord has been broken completely in two, the proper splicing procedure is simple (see below).

SPLICES AND TAPS

Joining the ends of two separate wires is called a splice. When a wire is joined at right angles to another continuous wire, it is called a tap.

The first step is to remove insulation from the wire about 1 to 1½ inches from the end. When using a knife for this removal, cut a slant as if you were sharpening a pencil, but be careful not to cut or even nick the wire under the insulation. With wire strippers, insert wire into the hole or opening of the correct gauge, close the strippers and pull the insulation off.

After peeling back the insulation, use your soldering iron or torch and some resin core solder to tin the exposed surface of the wire. This is not actually necessary with solid wire, but it is important with wire that is stranded. The tinning process does two things. First, it holds the strands together. Second, it provides a secure and permanently clean

FIG. 3-13 Using a wire stripper.

end. (Copper will eventually corrode. Solder won't.) The end result makes for a better electrical contact in the finished joint.

Heat the wire with the iron, and apply solder. (Do *not* heat the solder to let it drip onto the wire. The wire, not the soldering tool, should melt the solder.) Shake off any excess, leaving the exposed end shiny and bright. If you've done the job correctly, the tinning will look like nothing more than a change of color. You shouldn't be able to see any blobs or other obvious signs of the soldering.

To make a tap joint along a continuous wire, peel away about 2 inches of insulation from the wire to be tapped, and about 1½ inches from the end of the connecting wire. Securely wrap the end of the tap wire around the continuous wire and carefully solder the connection. Then wrap the entire joint with electrical tape to reinsulate the connection, leaving three wires coming from the one joint.

To splice the ends of two separate wires, strip the insulation, pre-tin, and then twist the ends to be connected tightly. A solderless cap may be screwed on, or the twisted wires may be soldered and then covered with electrical tape.

However you finally join the wires (solder, tape, etc.), the splice or tap itself must be physically strong. A gentle tug before completing the job will tell you. Don't assume that the solder is strong enough to hold the wires together. The only time you don't need to worry about this is

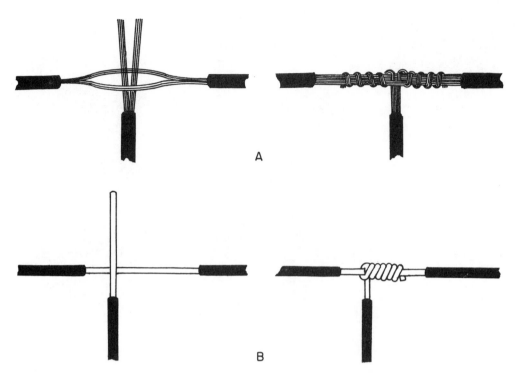

FIG. 3-14 A. Wire tap method for standard wire: *left*, intertwine wires; *right*. twist tightly to make a physically secure tap. B. Wire tap method for solid wire: *left*, strip the end of a wire, and the spot to be tapped; *right*, twist the tapping wire slightly to secure.

when you are using a solderless screw cap, since this automatically twists the wires together for a secure splice.

To fasten the wire to a screw terminal, bend the end of the wire into a hook or loop. Attach the hook *in the direction the screw will turn when tightened*. This will hold the wire tightly under the head of the screw. If the loop is inserted the other way, tightening the screw may cause the wire to slip out from under the head.

In instances where the cord was broken close to the appliance end instead of the plug end, use the screwdriver, nut driver, pliers, or Phillips head driver to remove any screws holding the case together. You must take off any plate or covering so that you can find where this other end of the cord is attached inside the appliance. Generally, one wire will connect to one side of the appliance on/off switch, and the other wire of the power cord will attach to another point (or terminal) inside the unit. Removing the short part of the damaged cord, stripping the wire ends, and reattaching the good longer part of the cord will give you practice in unsoldering and resoldering electrical wiring.

To unsolder the old connection, hold your soldering iron or gun in

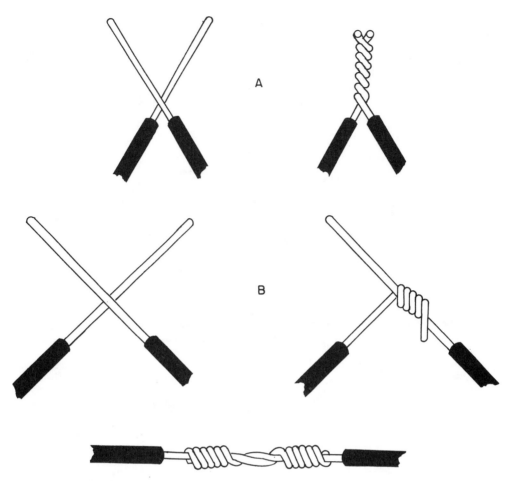

FIG. 3-15 Splicing methods. A. *Left*, remove insulation; *right*, twist wires tightly. B. *Left*, strip and cross wires; *right*, wrap downward on one side, then do the other side; *bottom*, completed joint is neat and strong.

one hand and the desoldering tool in the other. Make sure the appliance is firmly clamped to the table or workbench, leaving both hands free to work on the unit without knocking it to the floor.

The soldering iron has to be held directly to the soldered contact point where the wire is attached long enough for the solder to completely melt. As the solder heats, it will become shiny and fluid. Place the sucking end of the desoldering tool directly over the terminal you are heating, with the iron still in place, and suck off the melted solder. You may have to repeat the process two or three times to get rid of all the solder and make the connection loose. Then use your needle-nose pliers to pull the wire loose from the terminal.

If the wire has been twisted tightly to attach it to the terminal before it was soldered, it may be necessary to melt the solder with the iron in

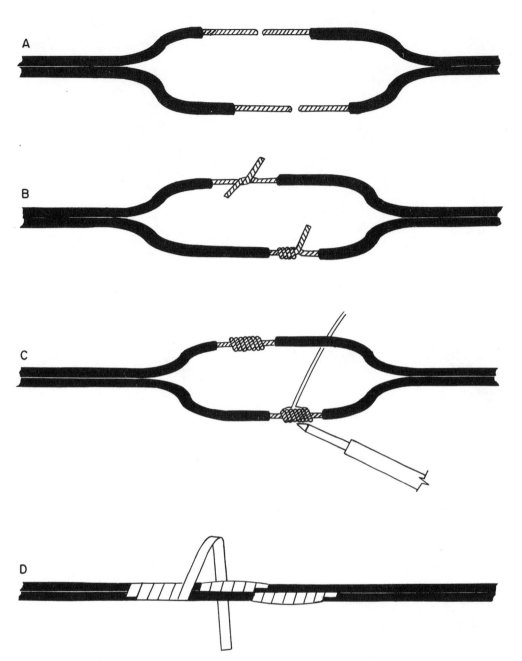

FIG. 3-16 A double splice. A. Remove insulation and stagger the wire ends. B. Wrap wires tightly. C. Apply solder to both joins. D. Wrap the finished double join with insulating tape.

one hand, and use the pliers in the other hand to wiggle and pull the wire loose from its connection. The appliance must be firmly clamped in place while the desoldering/disconnecting attempt is being made.

When the break or insulation damage is somewhere near the middle of the cord, the wire ends should be stripped and tinned (lightly coated with solder). Apply the soldering iron directly to the exposed end of the wire, and then apply a length of resin-core solder to the side of the wire away from the iron and hold in place until the solder begins to flux (melt). When the wire begins to take on the shiny, silvery appearance of a coated piece of metal, take away the iron and shake off any excess solder. This pre-tinning process will make it easier to solder the wire end to any other connecting point, and will serve as a binding agent to keep the individual strands of twisted wire from fraying and causing an accidental short.

After pre-tinning all four wire tips — two on each cord end (six tips if the cord is for a grounded, three-prong electrical connection) — take one of the wires from the longest end and twist it together with an end from the shorter piece of cord. In case the wires of the cord are covered with insulation of different colors, match the colors you are twisting together — black to black, white to white, green to green, red to red, etc. If the wires are of the same color, then use the VOM to check for continuity between one end of the wire still attached to the appliance and to the chassis or case of the appliance. This wire should go to the wire on the plug end of the cord that attaches to the widest prong. In two-wire electrical outlets, the wide prong, feeding to the wider left-hand opening of the wall outlet, is considered to be the ground side of the power cord.

Take your soldering iron or gun and apply heat and resin-core solder to the twisted wire connection. After the soldered connection has cooled, use wire cutters or diagonals to trim off any excess lengths of wire extending beyond the soldered joint. Then take your roll of electrical tape and wind a length around the spliced connection, making sure not to leave any wire exposed. The tape is to replace the insulation peeled off with a knife before the splice was made. It is not necessary to wind a lot of tape around the soldered connection, since the insulating quality of electrical tape is very good. Two or three layers of tape should be enough.

Repeat the process with the second wire from both the long and short end of the cord, again matching colors of the wire or wire insulation. The trick is to be sure that no point of the soldered connection on one wire makes physical contact with an exposed wire from other parts of the splice. If you have a three-wire connection, splice the third ground after you have used tape to reinsulate your second soldered joint.

Use the VOM to check the quality of your splice after completion. You should show continuity from one prong to one wire at the other end of the cord, and no continuity or an open circuit from that wire end to any other prong.

PLUGS

When the break in the cord has occurred very close to the plug, or at the other end of the cord near the place where it enters the actual appliance,

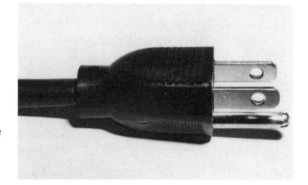

FIG. 3-17 This kind of plug is a permanent part of the cord. If the plug is bad, you'll have to cut it off and attach a new plug to the clipped cord.

FIG. 3-18 Electrical plugs.

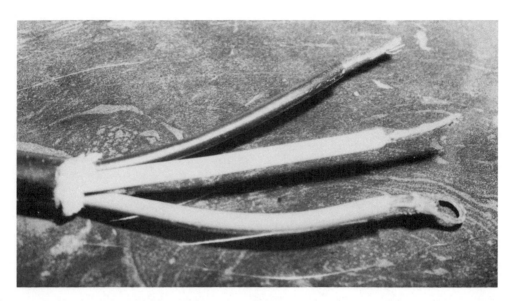

FIG. 3-19 A 3-wire cord, ready for its new plug.

you can discard the shorter part of the cord and reattach the longer portion.

On most modern equipment, the plug is solidly attached to the end of the wire and cannot be removed and resoldered. In this case, purchase a standard replacement plug from any hardware store or supermarket. Dress the end of wire from the longer cord at the place where it was broken; that is, make the exposed ending clean and neat. Use a knife or wire cutters to clip off the damaged end neatly. Make the cut at a slight angle or on the diagonal (about 30 to 45 degrees).

Most general replacement plugs are the slip-on and clamp, no-soldering type. A clip on the side of some plugs has to be opened. Push the cut end of the cord firmly into the new plug, and close the clamp firmly. Inside the clamping part are two needle-like contacts that will puncture the insulation on the ends of the cut wire and establish a connection between each wire and one of the prongs of the plug.

Heavier duty appliances (which need higher values of current to operate) will use cords with thicker wire that can carry this current without overheating. Many of these cords cannot use the slip-on type of replacement plug, but require a connector with screws and a mounting plate for attaching the cord. In this case, you will have to strip the ends of the wire before attaching the plug. That means you'll have to remove about 1 inch of the insulation from the end of each wire in the cord.

If you have a solid rubber or plastic-coated cord that seems to be only one wire, use a sharp knife to carefully peel away the outer layer of

FIG. 3-20 Test the new plug and cord for continuity.

insulation. This will expose the two (and often three) small wires inside the cord, each individually wrapped again with insulation. The three-wire cords are for appliances that must be grounded for safe operation. This means that the case or chassis of the appliance must be electrically connected with earth ground in order to prevent accidental shocks to the operator. Such appliances should only be used when plugged into a three-hole, electrically grounded outlet. The replacement plug should be a three-prong plug. Figure 3-19 shows a three-wire cord, properly stripped for the attachment of a new plug.

Twist the exposed ends of each wire, making sure you keep the strands of one wire from touching or twisting with the strands from either of the other wires. (Tinning the ends will help.) Push the cord up through the hole at the end of the plug away from the prongs. Wrap each of the exposed wire ends around one of the three contact screws leading to the individual prongs. Again, make sure no frayed strands are left that may touch from one wire to another. Tighten the screws. This procedure is best practiced on a throwaway piece of electrical cord that is open at both ends. You can use the VOM to test your work before trying to repair an actual appliance.

Chapter 4
Small Appliance Repair Basics

Diagnostics and servicing, whether on a small appliance, large appliance, your car, home computer or transistor radio, is always about the same. It's a basic 1 − 2 − 3 approach.

1. Diagnose − Test the appliance to see what is wrong in general. Find the general symptoms (what works, what doesn't).

2. Pinpoint − Isolate the problem to the particular part causing the trouble.

3. Repair/Replace − After you've located the specific cause of the problem, repair or replace the bad part(s) and put the appliance back into service.

BASIC DIAGNOSIS

Diagnosis is nothing more than a process of elimination. There are only so many things that can go wrong. By eliminating those things that *are* functional, you'll find what *isn't*.

The diagnostic procedure begins with the simple and obvious, then goes to the more complex. Much of the time the problem will be simple and obvious, so why waste time and energy, and risk damaging the appliance by tearing it apart, when you don't have to?

Much of this process of elimination involves an "either-or" approach. For example, either power is getting to the appliance or it's not. In one quick and easy step you've identified either the appliance as the problem, or everything else coming to it.

At this point, if there is no power to the outlet, you'll know that taking the appliance apart will be a waste of time. The trouble probably isn't with the appliance. (If it is, the appliance has a major short circuit that is blowing the fuse or popping the breaker. And this in turn tells you something very important—you're probably looking for a bare or burned wire.)

If there is no power to the outlet, then either the problem is between the breaker and the outlet, or it's between the breaker and the power company's primary. This might sound oversimplified, but I knew a person who spent several hours ripping apart a nonfunctioning blender, and damaged his kitchen counter, only to find out that someone had crashed into a power pole and all the power in the neighborhood was out. (And then he couldn't figure out how to put the blender back together!)

If you have power anywhere in the house but none at that outlet, then you already know that the problem is within a specific circuit. Replace the fuse in that circuit, or reset the breaker. Now try the outlet again. If the appliance goes out again, one more quick test will tell you whether the trouble is in the appliance (a short circuit) or in the wiring

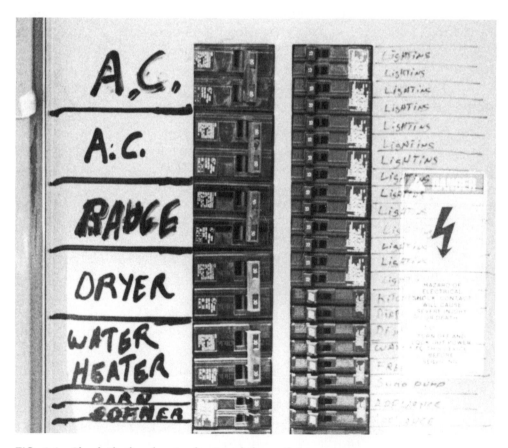

FIG. 4-1 Check the breaker (or fuse) and the outlet.

Chilton's Guide to Small Appliance Repair/Maintenance
SMALL APPLIANCE REPAIR BASICS

(also a short circuit). Reset the circuit again and try the same outlet with another appliance. (Also don't forget that the whole problem might be that you have too many things on that circuit. If the circuit is meant to supply a maximum of 15 amps, and you're trying to draw 20 amps from it, it will blow again and again until you remove the overload.)

Now if the circuit blows when you plug in your alternate appliance, you know that there is a short somewhere in the wiring between the breaker and the outlet. You can eliminate the breaker and the outlet itself easily. First visually inspect both (with the power off). You might see signs of a burned wire, which means that arcing has taken place. This is usually a sign of either a bad component or a loose wire. (Chapter 3 will give you more details on how to check/replace the outlet and fuse or breaker.)

If the circuit is not overloaded, and if only the suspected appliance is causing the breaker to let go, you should look for a short circuit in the appliance.

Short circuits, whether within the appliance or in the wiring of your home, must be located and repaired. This is both for proper diagnosis and repair of the appliance, and for safety. You will not be able to further test the device until you locate and cure that short, or until you can fix the device to the point where you can plug it in and turn it on without again blowing a fuse or tripping a breaker.

On the other hand, if there *is* power to the outlet, and if other appliances work just fine in that outlet, then you know that the appliance has indeed something wrong with it.

Within a couple of minutes you'll have eliminated at least half of the possibilities and can then concentrate on where the problem really is. If the problem is definitely in the household wiring, go back to Chapter 3. If it's in the appliance, continue reading.

The next step involves thinking and asking yourself a number of questions. Since you know that appliance, you probably know the answers to many of the critical questions that make diagnosis easier. Once again, begin with the simple and then go to the more complex.

Were you using the appliance when it went bad, or was the appliance working perfectly last time and failed to work the next time you went to use it? Has it ever worked? Has that particular function ever worked? Have you checked all the obvious things, such as the plug? (A large percentage of "malfunctions" are something obvious, or even silly, with nothing being wrong at all.) Does the person who first reported the malfunction really know how to operate the appliance?

If the appliance is new, you must answer "I don't know" to the question "Has it ever worked?" Even if it has been around for a while, it's pos-

Initial Diagnostics Checklist

1. *Have I read the instructions?*

2. *Has the appliance ever worked? Has that function ever worked?*

3. *Is ANY part of the appliance working?*

4. *Is power getting to the outlet?*

5. *Is power getting to the appliance?*

sible that you're trying out a function that is new for you. Have you read the instructions? It could be that you're not doing the right things to get that function to operate. It's even possible that you're trying to get the appliance to do something it can't do.

If you've had the appliance for some time, and have used it successfully all that time, then it's less likely—but still possible—that you're doing something wrong. But you still can't assume that someone else (who says it's not working) knows how to operate it.

Especially when someone else told you that an appliance isn't working, you must confirm that the appliance *is* in fact nonfunctional. Then examine the unit while still trying to make it operate, so you can begin to diagnose the real problem. Now you begin to take note of the symptoms.

Did it overheat after long use and then malfunction? (After it's cool, if it seems to be working all right, it could indicate a lubrication problem, or electronic components that are wearing out.) Did a simple mechanical jar, made while moving the appliance from its normal working space to the shop, jiggle a loose part back into place?

If the appliance has a heating element that is failing to even warm up, don't immediately jump to the conclusion that a thermostat or the heating element itself has gone bad. Begin the process of elimination. Think of the basic rule of electricity (it must have a complete path) and use your common sense. You'll know what parts to test.

With the heating element example, first there must be electricity available. At this point you should have already eliminated all things outside the appliance and up to the outlet. Now you can take one more step. Is power getting *to* the appliance? If that appliance has an "on" indicator and this fails to light, that is a sign that perhaps power is not get-

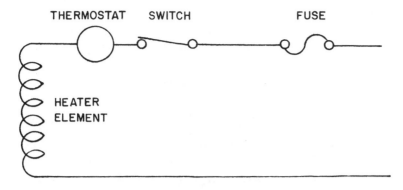

FIG. 4-2 Schematic of a heating element.

ting to the appliance. The trouble could lie in the unit's power cord, power switch, or in any of the wiring or electrical controlling/switching devices that feed the heating element.

After you've determined that the fault is definitely with the appliance, it's time to find out just exactly what is wrong inside. As in the example above, the first part of diagnostics will help to eliminate many of the parts so you don't waste time on them.

PINPOINT AND ISOLATE THE FAULT

Step 2 in the diagnostics process is to pinpoint the problem. For this, you may have to take the appliance apart in order to test and visually inspect what is happening inside the casing. Most of the time, how to take the appliance apart is obvious. Some tips will be given in the next section of this chapter.

Whatever malfunction you are able to verify by testing the appliance for yourself, try to use common sense and your knowledge of the work normally accomplished by the device in order to think of all the wiring and parts that feed or that conduct electricity up to the part that is not doing its job. Use a notepad and jot down all the things you believe might be preventing the malfunctioning part from working.

If the motor doesn't turn, the actual problem might include power not getting to the outlet, a faulty power plug or cord, a faulty on/off switch or other function (such as speed) switch, an open wire inside the unit, motor binding due to lack of lubrication or a bad motor bearing, the windings of the motor itself might be shorted or open, or the blades or motor driveshaft could be physically bent and/or jammed.

Once again, go to the easy and obvious first. Use your senses. You might be able to see what is wrong. A wire might be burned, broken or loose. A belt may have fallen off. Look around carefully, both before you remove any panels or the casing, and afterwards. (Be sure that the power is off and the appliance unplugged before attempting to get inside!)

Continuing with the motor example, listen for a humming or vibration. If you hear it, you can probably eliminate (at least for now) those electrical things outside the motor. Some power is getting to the motor, or it wouldn't be making noise.

Shut off the power and see if you can find something mechanical at fault. Is the motor secure in its mount, or is it flopping loose? If the motor drives the appliance with belts, are the belts tight enough (but not too tight)? Are the belts even there? If the motor is gear driven, are the gears meshing properly? Are all the teeth of each gear intact?

Manually try to turn the shaft, if possible. Any binding should be fairly obvious. It should also be obvious if the binding involves the motor shaft, or whatever part of the appliance the motor is trying to spin.

If you keep the functions of the appliance and of the various parts in mind, you should be able to quickly and efficiently track down the source of the trouble.

The three overall functions of a household appliance are electrical, mechanical, and plumbing. A problem in the third category will generally be obvious, since it will usually leave a puddle of water. A coffee maker, for example, with a loose or broken hose inside might make a half pot of coffee — with the other half to be found on the counter, and possibly inside the coffee maker causing it to short out and blow a fuse.

Electrical problems are easy to track down by just keeping the "complete path" rule in mind. If electricity is getting to a certain component, but that component isn't working, it's highly likely that the component is at fault. If current isn't getting there in the first place, then the problem will be found ahead of that component.

To test a component for continuity, you'll usually have to remove the component. At very least you'll have to remove at least one of the wires (otherwise you may get a false reading and end up actually testing something elsewhere in the appliance).

A heating element, for example, should have a very low resistance reading if it is whole and sound. If it is starting to go, the resistance will be fairly high. If a wire within the element has burned through, the continuity testing will show an infinite resistance, indicating a broken path. If so, it is time to replace the element.

Mechanical problems can often be tested manually. A motor shaft can be turned. A button can be pushed. They can also be inspected visually for excessive wear, metal flakes, broken or loose parts, and an odd appearance in general.

Sometimes a component has more than one action. A button has a mechanical motion, and this movement may control something electrically. Pushing that button makes a physical, and consequently electrical, contact. In such cases, testing might include testing with a VOM and testing with your fingers.

DISASSEMBLY

In disassembly, your first concern will be to find out exactly how the appliance is put together physically. Even if you have an instruction or service manual for the device, most do not detail the correct way to get inside. Other books, and even the owner's manuals, might say a lot about what you can do once inside, but they may not tell you how to "break and enter" in the first place.

Keep in mind that any and all warranties might be voided with the first turn of the first screw.

Always switch off the power and unplug the unit. Even then, keep in mind that there might be charge-holding devices inside (such as capacitors) that can still be dangerous. Move carefully and keep your eyes open.

Now carefully inspect the case and the obvious holding screws. Many screws or bolts go inside the unit and hook up or fasten to some-

FIG. 4-3 Visually inspect the appliance cover(s) to find the screws, bolts, and other fasteners. Be careful not to remove those fasteners that hold internal parts.

Chilton's Guide to Small Appliance Repair/Maintenance
SMALL APPLIANCE REPAIR BASICS

thing inside the case, too. Usually these are obvious, but not always. If you're in too much of a hurry when you take the appliance apart, a spring might pop loose and flip across the room. You could lose a part, and not be able to discover exactly where it was supposed to go anyway. Use care in removing any screw or clamp. As a general rule of thumb, screws toward the outer perimeter merely hold the case in place, while screws toward the center might be holding something else.

When in doubt, first slightly loosen the screws you *think* hold the case and nothing else. With the screws loose but not removed, you should be able to tell if something else is coming loose inside, or if just the case itself is being released.

Once all visible screw heads have been removed or loosened, carefully try to lift the covering plate or appliance case away from the unit. If it does not move, run your fingers around the edges to see where the cover or case is still attached. You may discover that you have to remove a function knob or a switch before the case can come off. *Never* force the case or cover off.

In these days of plastic mass manufacturing, many cases that appear to be sealed single pieces are actually snap-apart units. To find the "secret" release(s) is a matter of probing and prying.

If all else fails, and if the unit is a throwaway if you can't fix it yourself, then — *and only then* — apply force. Expect to physically break the case, and possibly destroy the appliance. At least you may find out how to open the next one without breaking it. There's a remote possibility that you will be able to fix the appliance, and then repair the broken case before reassembling.

In the disassembly process, use your notepad to draw diagrams of exactly what small parts go where, in what order they are mounted on a shaft, and what color of wire goes to what plug or connection.

Once the unit is apart, it is quite possible that you will find the malfunction has been caused by something easily repairable. A wire could have broken or could have pulled away from its terminal. If you are sure what terminal the loose wire is supposed to have connected to, then resolder it in place.

Be very careful here after reassembly and plugging the appliance back into a wall outlet. If you soldered the loose wire to the *wrong* terminal, you may get a short circuit. Such a mistake could cause a major burnout and a major problem, when the original loose wire was a minor problem — if only you had hooked it back up in the right place. It can also be a potentially deadly mistake.

For this reason, if you find a wire disconnected inside and are not absolutely sure (either from diagrams in the user's manual or from the

FIG. 4-4 Once inside, carefully inspect for any signs of obvious damage.

broken remains of the connection on the terminal) then consult a service person at the shop where you bought the appliance to be sure "point X" is precisely where the loose wire is supposed to be soldered.

(There are some very rare instances when a loose wire inside is supposed to be loose. The manufacturer needed a two-wire cable to hook up part A to part B, but was able to mass-purchase a bunch of three-wire cables at a warehouse fire sale somewhere. The maker would use two of the wires in the three-wire cable to manufacture the appliance, clipping or taping the spare wire somewhere out of place. Taking the unit out of its case may have loosened that unused third wire — so don't just guess that it is supposed to be resoldered to some point. Check with a technician to be sure.)

The VOM can be a major help in your pinpoint work. You can use the resistance scale to test continuity or to see if a connection where the solder looks old and cracked is still making a good electrical contact. Sometimes using a VOM to test what appears to be a bad component will reveal that it does *not* need to be replaced.

At this point, if you have still not found the broken part or wire in the appliance, place it carefully on the workbench so that no exposed wires are touching each other or making contact with outside metallic

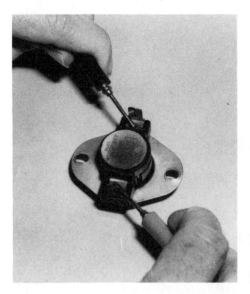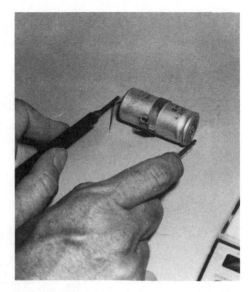

FIG. 4-5 Using a VOM to test the continuity of a component.

surfaces. Plug the unit into a wall outlet, turn the power switch on, and use the voltage range (ac voltage) to be sure the electrical pressure is getting to the switch or to a malfunctioning motor or heating element.

If the appliance uses a dc motor instead of an ac motor, it will have a power supply circuit board for internally converting the wall outlet ac to dc. The trouble could be in this power supply. (See the next section in this chapter.)

Remember to set the VOM to its highest voltage range before applying power, and then reduce the range switch if necessary to read the voltage accurately. For most home service work, however, measuring exact voltage values will not be necessary. If voltage pressure of any value shows on the meter when the probes are placed across two points that should indicate a voltage, then the circuit is probably working all right. For this reason, some shop people use an inexpensive ac/dc testing light bulb instead of a VOM for voltage checks.

The diagnostic and pinpointing steps involve a lot of physical inspection and some solid mental work. "What *could* be wrong to cause the unit to do this?" Why waste your time testing the motor when the symptoms show a problem with a heating element only? (Remember the old adage "If it's not broken, don't fix it.")

With over half of your malfunctioning small appliances, the pinpoint steps will lead you definitely to the broken or bad part. You will be able to *see* what has gone wrong, and will be able to either repair, resolder, or replace the defective component. Once you have done this, reassembled the appliance, and then tested it to see that it is again working properly, you have achieved your first money-saving, quality appliance repair job!

If you've reached the point where you *think* you know which part is bad (because everything else seems to be okay), then it's best to read further into the chapters on individual parts testing before spending bucks for a new part (that may actually not be required). If the part checks good, you'll have to repeat these preliminary diagnose/pinpoint steps, and will probably discover that you missed something. You assumed a part or wire connection was working, and it really wasn't.

POWER SUPPLIES

Some appliances use microchips and other electronic circuitry that requires voltage other than what comes from the wall outlet. If your appliance is "computerized," this is definitely the case.

To operate, those circuits generally need dc voltage of between 5 and 24 volts. Since this doesn't come from the wall outlet, something must be built into the appliance to convert that 120 vac into the needed value of dc. This something is a power supply.

There are three basic parts to a power supply: the transformer, the rectifier, and the filter(s).

The transformer takes the incoming 120 vac and drops it to near the final value(s) needed. This happens because of a difference in the number of windings between the primary and the secondary sides of the transformer. If that ratio is, for example, 10:1, then the 120 vac will be dropped to 12 vac. (This is an oversimplification, but gives you the general idea.)

The size of the transformer depends on how much current it has to

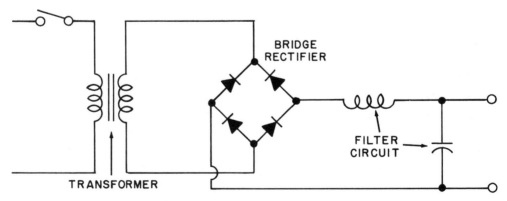

FIG. 4-6 Schematic of a simple power supply.

FIG. 4-7 A power supply transformer.

FIG. 4-8 A bridge rectifier.

handle. If the components needing dc voltage (or even a different value of ac voltage) draw a large current, then the transformer will be large. If very little current is required, the transformer can be small.

After the voltage has been changed to the proper value, a rectifier is used to convert the ac into dc. A rectifier allows current to flow in one direction only, while blocking the flow in the other. This is done with a component called a diode.

The usual rectifier contains multiple diodes. If just one diode is used, half of the incoming current passes through; the other half is "thrown away." Current flows only half of the time. By using two diodes, when one diode is blocking half of the flow, the other diode takes over and passes it along, still in the same, single direction. Many power supplies use what is called a bridge rectifier, which is four diodes connected in a square pattern. Most often this bridge rectifier is a single block.

Once the current comes out of the rectifier, it is direct current only in the respect that it flows in only one direction. It still varies in value from zero to (for example) 12.

A filter capacitor takes care of the rest. As the current flows into the capacitor, it fills up and stores the energy. It then releases it in a steady flow. The end result is virtually pure dc of a steady value that can be used by the other components.

Testing a power supply is easy. First find where the ac comes in. Set your VOM to read ac in the proper range (usually the 120 vac line voltage) and see if ac is getting to the power supply. If it is, find where voltage leaves the power supply. Most of the time, the power supply output will be dc at between 3 and 24 volts. Set your meter to read at least this much and test the output(s).

If power is getting *to* the supply but is not coming out, the power supply is bad and has to be replaced. If it is coming out, or power isn't getting to the power supply in the first place, then the problem is elsewhere.

BUYING PARTS

In many instances, you will discover that the problem with a nonfunctioning appliance can be cured by cleaning, lubricating, or by physically bending or twisting a misshapen part or bracket back into its correct position. To effect repairs in other cases, you will find it necessary to purchase a replacement part. Sometimes parts availability is a major problem, but luckily this is seldom a troublesome factor in appliance work — unless the make or model of appliance you are working on has been discontinued by the manufacturer quite a few years before.

Once you have determined what replacement part is required, make a note of the brand name and model number of the appliance. This is generally listed on a plate somewhere on the outside of the unit.

Either go to the store where the device was initially purchased, or consult the Yellow Pages of the telephone book in order to obtain the name of the main dealer in your area for that brand of appliances. The service shop of that store will have the part on hand or should be able to order it for you.

Also listed in the Yellow Pages are likely to be stores that carry appliance parts. (After all, the professional service people have to go somewhere to get their parts.) Chances are very good that you'll find more than one in any medium-size city.

Some appliance components are available at general hardware, electrical, or electronic supply outlets. In most cases, if a part is generally available, it is less expensive to pick it up from the counter of one of these stores than to order the replacement from the original manufacturer.

Occasionally you will run across an electric motor, a heating element, a thermostatic control device, or some other commonly used component that is not stocked on the counter of a general supply store, but which may be ordered directly from the part manufacturer rather than from the maker of the appliance. The motor of a Brand XXX vacuum cleaner, for example, may have been supplied to the Brand XXX company by International Electric Company, with the IEC home address stamped right on the motor casing. Write for an IEC catalog, or call to see if there is a local dealer. Buying direct from IEC may again save you the normal mark-up charged by Brand XXX if you buy the replacement from that manufacturer.

Many mechanical parts that have broken or that have been damaged in other ways can be replaced by duplicates you fashion yourself in your home workshop. Or the part could be fashioned by a local machine/tool shop for you at a reasonable price if the component is a fairly simple one. This could prove necessary if the part is not available to buy and if the price of the custom-made part is less than the price of a new appliance.

It's an excellent idea to keep your old nonrepairable appliances as a stock of replacements for newer units when they, in turn, go bad. This is a good reason for sticking to the same brand and model of each particular appliance you use (once you find one that is completely satisfactory in its service).

For small appliances this is easy. If an old Mr. Coffee maker develops a bad switch and the unit is so old you'd really rather have a new one, don't throw away the bad coffee making machine. Some day your new Mr. Coffee unit may encounter a burnt-out heating element, and you'll have a quick repair part right out in the garage or workshop.

Even when switching model or brands in a new appliance purchase, some parts may still be common. You can disassemble the broken unit and store all the individual parts in a conveniently marked and labeled parts bin or storage cabinet. A switch, thermostat, bracket, or heating element may work as a future repair part not only for the new appliance you bought, but also in some other appliance that just happens to require the same type of component.

Chapter 5
Electric Motors

Almost every appliance in and around the house has either a heating element or a motor or both. The common solution when a motor goes out is to replace the motor. Sometimes this is the only solution available, especially with small appliances. However, there are some very good reasons for not using this as the *only* solution.

First, you may be able to restore the motor to operating condition — often quickly and easily. You may find that what seems to be a major problem is something simple; or perhaps the problem can be fixed by installing a new part (instead of a whole new motor).

Second, even if you do buy a new motor in order to get an appliance back into working condition, the old motor can serve some useful purposes. Unless the old motor has a trade-in value, take a few moments to break down the old one. Once again, you might find that the terrible problem is something fixable, which will give you a spare motor for next time. If it can't be fixed, perhaps the good parts in the old motor can be salvaged, thus adding to your personal stock of parts. (This is especially important when spare parts aren't normally available.)

Perhaps most important is that the old motor, fixable or not, can serve as a valuable training aid. Sooner or later you're going to come across another use for that knowledge. What you learn about the construction and inner workings of the motor can save you time and money in the future.

Even with all of this aside, understanding how a motor works and does its job in the appliance can help tremendously when it comes time to swap out a faulty motor.

MOTOR BASICS

A motor changes electrical energy into mechanical energy. To oversimplify: plug it in and it spins. The amount of force in that spinning is called torque. The power of a motor is generally given in horsepower. Very small motors often have no meaningful rating of this nature. But whether rated or not, and no matter how the motor is rated, you should always try to get an exact match when replacing a motor. (Quite often, if the replacement motor isn't an exact match, it won't even fit.)

Don't worry if you don't understand completely how motors work. Your goal isn't to design a motor, but to diagnose and cure problems. In brief, motors work on the principle of magnetism. Like magnetic poles repel each other, while unlike poles attract. If you bring the south pole of one magnet near the south pole of another, the two will repel each other. If you set up a number of magnets in a wheel with a ring of magnets around this, and if you could constantly shift the polarity from south to north to south, etc., you'd have a motor.

Fortunately, electromagnetic forces will do this for you. Electricity flowing in a wire generates a magnetic field, causing the wire to become a "magnet" temporarily. By using polarity switching (either mechanically or by using ac voltage, which constantly changes anyway), the motor spins merrily away to get the job done.

Motors have various types of voltage connections to their internal

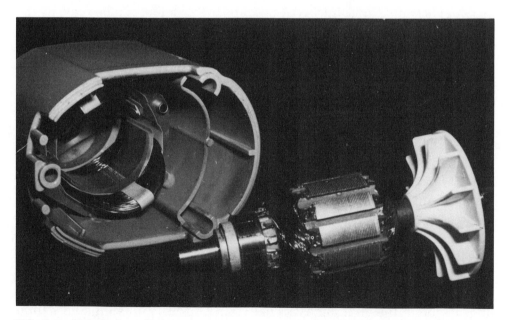

FIG. 5-1 A typical appliance motor.

coil windings. *Series-wound* motors have field windings connected to the brushes of the armature so that current passes through the field windings before reaching the armature (mounted on the rotating shaft of the motor). These motors are used to move heavy loads that have to be started slowly (as with a garage door opener, for example). Series-wound dc motors must be operated always under load, to prevent the speed from building up to the point where the motor may self-destruct.

Shunt-wound motors have a parallel wiring setup to provide current to field windings independently from that supplied to the armature coils. Shunt-wound motors are used where constant speed is necessary under varying load conditions (in good quality electric shavers, blenders, etc.).

Compound-wound motors incorporate both series and shunt armature windings. These motors can handle heavy starting loads and can draw energy from the centrifical force of a spinning flywheel during peak load conditions. The flywheel also absorbs extra energy from the motor after peak demands pass, helping to maintain speed of the motor and to increase the working life of the device.

The electric motor is designed to provide a turning force (torque) to accomplish some designed function, such as opening a can, opening a garage door, providing a breeze, crunching your trash, or playing your audio and video tapes. With such a variety of things employing electric motors, it is not surprising that they come in a variety of forms.

There are three main groupings of electric motors, each designed for a specific type of power source: direct current (dc), single-phase alternating current (ac), and polyphase alternating current (ac). The first and second are those commonly found in homes, while the third is more likely to be found in commercial applications.

AC MOTORS

There are a large number of ac motor types, each with special characteristics to meet the needs of various appliances. The most commonly used types of ac motors include: universal, shaded pole, capacitance start, split phase, PSC (permanent-split-capacitance), and synchronous. These motors all share the common trait of requiring ac to operate, with the exception that the universal motor can be designed to operate on direct current as well.

The universal motor is aptly named, since it can be used to handle so many different jobs and can operate on either ac or dc, depending on the way the appliance (and the motor) is set up.

Shaded-pole motors can be used when the amount of torque needed

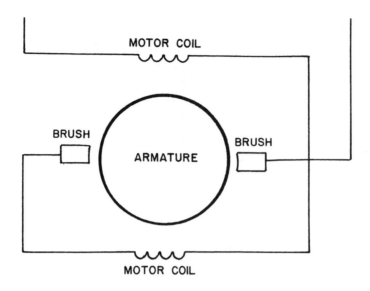

FIG. 5-2   Schematic of a universal motor.

is relatively small. This motor has four stator poles, with a band of copper somewhere on the poles. The copper band "shades" the magnetism, thus causing the same pole to have more than one magnetic field, which in turn causes that magnetic field to rotate, which in turn causes the motor to rotate.

Watch the lights in your house when a major appliance, such as a large air conditioner, starts up. They tend to dim for a moment. The reason is because it takes more current to get a motor started than it does to keep it going. This is hard on the electrical supply, and hard on the motor. Nicely enough, there is a solution. A capacitor stores energy. When the motor starts, the capacitor releases its energy, giving the motor the needed boost. Capacitance-start, split-phase, and PSC motors are rarely found in small appliances.

Split-phase motors can also be used in large appliances. Two windings are used instead of just one, with a centrifugal switch to shift from one to the other. The start-up winding brings the motor up to about 80% of operating speed. The centrifugal switch then cuts out the start-up winding and brings in the main winding. Then the remaining 20% of operating speed is gained and held.

PSC motors are generally found in larger appliances such as air conditioners and other devices requiring large amounts of torque. These motors are like a combination of capacitor start and split phase motors. In a PSC motor, the capacitor is permanently connected in series with the motor, which makes the start-up winding do double duty. At startup,

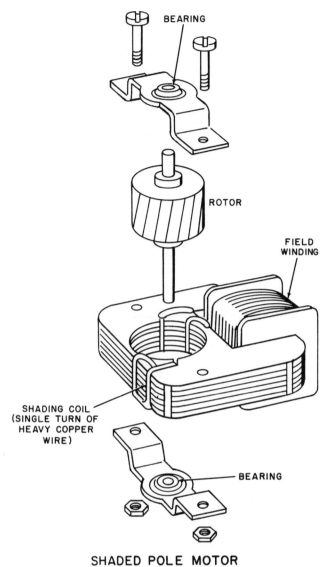

BEARING

ROTOR

FIELD
WINDING

SHADING COIL
(SINGLE TURN OF
HEAVY COPPER
WIRE)

BEARING

SHADED POLE MOTOR

FIG. 5-3   Shaded pole motor. Reprinted by permission from *Reader's Digest Fix-It-Yourself Manual* (New York: Reader's Digest General Books, 1977).

this winding works in the same way as the startup winding of a split phase motor. As the motor reaches speed, the winding will work as the main winding.

All ac motors are *synchronous*; that is, they are automatically synchronized to the incoming 60-cycle power. This causes them to run at a designed speed. (The speed of a dc motor is determined by the amount of current flowing, since dc has no frequency.) Appliances sold in the Unit-

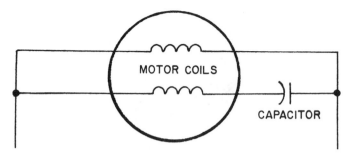

FIG. 5-4 Schematic of a capacitance start motor.

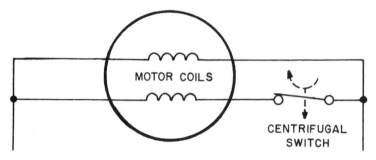

FIG. 5-5 Schematic of a split phase motor.

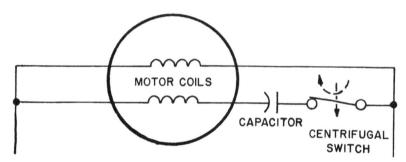

FIG. 5-6 Schematic of a permanent-split-capacitance motor.

ed States are designed to operate on the standard 60-cycle frequency supplied by power companies to the wall outlets of your home. The speed of a *hysterisis/synchronous* motor is dependent upon that 60-cycle frequency (and with the number of poles used in the manufacturing process).

For those interested, the basic formula is S = 120 x f/p, where S is the speed of the motor, f is the frequency of the line voltage (60 cps in the United States) and p is the number of poles in the motor.

The formula isn't exact. There are other forces applied against the motor, such as the voltage level and the load the motor is expected to handle. Most motors operate within about 5% of the speed resulting from the formula. And actual operating speed of most motors tends to vary across a fairly wide range.

There are times when this speed variation isn't sufficient for the job. There are times when an exact motor speed is essential. Many times when you hear someone talking about a synchronous motor, they are talking about a specially built motor that is designed to operate at this exact speed, and one that is relatively immune to variation.

As mentioned above, it doesn't really matter if you understand all the details of how each motor type works, or why a shaded-pole motor is used in one appliance while a PSC motor is used in another. You don't have to worry about recognizing one over another. The underlying principles are the same, regardless. So are the techniques of troubleshooting.

DC MOTORS

The other common motor found in and around the home is the dc motor. They are often found in toys and they are usually battery powered. However, there are numerous exceptions to this rule: dc motors are often found in audio and video cassette decks and other high-tech goodies. The most common form of dc motor uses an internal permanent magnet to provide the magnetic field in which the armature operates. Other forms of the dc motor use various methods of connecting the stator and armature to enable the motor to operate using a dc power source.

A convenient feature of dc motors is that the direction of rotation can be changed simply by changing the polarity of the current feeding it. This is how the motor direction is changed in order to reverse the movement of the device. The switching arrangement for the motor and connected device contains an internal polarity changing feed to the motor.

You will sometimes find compound-wound dc motors in fans and blowers and in heating and cooling household appliances. The speed may be controlled either by switches or by a rheostat that varies the revolutions per minute (rpm) over a wide range (by varying the current). The more common two- or three-speed manual switching arrangements incorporate a number of resistors that can be cut into or out of the fan/blower motor circuit at the discretion of the operator. When the switch is in a low-speed position, more resistance is added to the motor wiring circuit than when the intermediate speed is selected. Removing all resistance from the circuit by throwing the switch into another position puts the motor into high-speed operation.

RELATED PARTS

In air conditioning and other heavy-duty motor appliance operation, a magnetic clutch is often involved in order to adjust the driving power of the motor. You will also find these clutches on electric lawn mowers and in some vacuum cleaners. Magnetic clutches are generally in the form of an electromagnetic coil (solenoid) mounted between the drive shaft and the belt pulley.

Electric motors sometimes have accompanying starter voltage supplies, clutches, drive belts, gears, and operational solenoids — so many circuits and subcircuits that it becomes difficult for even professional service workers to keep abreast of new developments. Yet despite the involved circuitry and mechanical principles, only a few basic electrical/mechanical processes are involved in their operation. Once you have studied the basic operation of relays, solenoids, and electric motors, new adaptions are easy to handle when you encounter them.

In today's world of solid-state electronic improvements, many dc motors are directly mounted on a small circuit board that can contain a transistor or two, coils, capacitors, and even an IC (integrated circuit) chip. The electronic parts mounted on the board are part of the motor control circuit. They can help in determining the rotation speed of the shaft, they can be part of any designed overload protection devices, or they can supply the correct starting voltages when the unit is first turned on. Generally, the manufacturer will supply a replacement motor *only* as part and parcel of the entire mounting circuit board. In other words, the motor (a simple $10 dc unit) cannot be purchased unless the consumer buys a whole new motor board (for $25 to $100 or more).

In this case, you would be wise to look for the name of the motor manufacturer, as opposed to the appliance manufacturer. It would be less expensive to order the motor directly from the maker, without the mounting circuit board and components.

A difficulty arises due to modern "proprietary information laws." The motor could be a fairly common style made and marketed world-wide by ABC Motor Company. But when XYZ Appliance Company builds a trash compactor or an electric juicer using this motor, XYZ will contract ABC to supply 100,000 or so electric motors with a single variance from normal manufacturing specifications from all the other ABC motors. An internal capacitor or resistor may be added when ABC assembles the motors for XYZ. This means that if you buy a "stock" ABC motor as a replacement for an XYZ appliance, it won't work. And neither ABC nor XYZ will let you know what tiny modification has been made from one motor design to the other. In this case, you will have to order the replacement motor and the accompanying attached circuit board (which is still okay on your own appliance) directly from XYZ Applicance.

The fact that a motor in your appliance may be mounted on an independent motor circuit board means that other test steps should be taken before ordering a replacement motor or motor circuit board. Use the VOM to test continuity through the board. If you are familiar with basic transistor testing, remove leads and use your VOM (or a transistor checker) to see if an inexpensive transistor in the drive board is at fault, rather than the motor. Most important, use the voltage scale of the VOM to see that the DC value applied to the motor terminals is correct for proper operation (24 volts for a 24-vdc motor, etc.). If the voltage is abnormal, then the trouble is probably not with the motor. Instead, check individual components on the motor drive board, and test values from the circuit's power supply. (See the sections on testing and troubleshooting below.)

MOTOR MOUNTS AND BEARINGS

To successfully replace a motor, you have to know how to unmount the old motor and how to correctly remount the new one.

Some motors are designed to be held rigidly in place inside the appliance. Others are suspended on belts or rubber mounts, allowing them to vibrate rather freely. Be sure, when replacing a motor, that you allow it to be mounted the same way as the original. Don't attempt to clamp down a motor shaft that is supposed to be suspended loosely, and be sure to securely bolt or screw in place those motors intended to drive from rigid mounting brackets.

In efforts to keep manufacturing costs low and to maintain moderate prices for consumers, basic motor design for many low-priced small appliances has changed a lot over the past decade. Many low- and moderate-speed dc motors are now constructed with plastic or silicone/plastic bearings.

A bearing is the small lubricated ball on which a motor shaft rotates.

It eliminates friction as the motor spins to drive a pulley or belt or whatever. In old-fashioned kid's roller skates, it was easy to see the ring of metal ball bearings held in a row around a metal ring or collar. This collar was locked in place, with the wheel shaft protruding through it. The shaft would turn as the wheels rolled, rubbing against the freely rotating ball bearings.

At that time, most motor bearings were built in a similar manner. They would last for a long time, but required frequent lubrication.

Nowadays, in many sealed motor units, the bearings are plastic. They do not last as long. If the motor is a sealed, one-piece unit, when the bearings wear out the motor or the entire appliance is defective and ready for replacement. In such cases, you might as well do a little experimental test work. What you learn just might make such a motor salvageable, for use as a future replacement. And you might gain sufficient knowledge to be able to repair the next motor breakdown without having to buy a replacement.

Whenever a motor using plastic bearings is *not* sealed, make it a point during routine maintenance checks to test the motor shaft's "freedom of movement." If the drive shaft wiggles too much when side-to-side pressure is applied, then the bearings are becoming worn. Exact replacements will be available, at low cost, directly from the manufacturer.

When an appliance is purchased, the retailer normally provides an operator's manual and a replacement parts list. Maintain a single file somewhere in the house for *all* your appliance manuals and repair information. If you order replacement bearings when you first notice a mo-

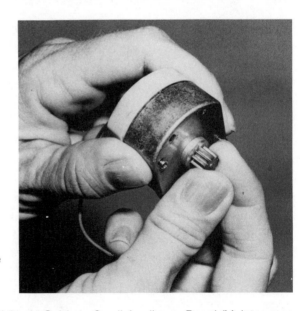

FIG. 5-7 Testing the motor's bearings can usually be accomplished by turning and wiggling the shaft and feeling for excessive play or excessive resistance to movement.

tor's rotation becoming "soft," then the new parts will probably arrive before the appliance suffers a complete breakdown.

Disassembly for replacement of bearings is fairly simple. Carefully study the way the unit was originally put together. Use the proper tool (screwdriver, allen wrench, nut-driver, etc.) to open the motor casing and remove the old bearings. Replace with the new part, being sure to align it exactly the same way as the original. Put the unit back together, simply reversing your disassembly steps.

Do not attempt to replace worn bearings with substitute "similar" bearings from your tool box. These fittings are critical for smooth, efficient, and trouble-free operation. Often the size variances are in thousandths, or even millionths, of an inch. What looks to be a bearing or bearing set of the same size could easily be wrong, and can damage the entire motor in mere minutes of operating time.

WHY DOES IT BREAK?

An electric motor usually becomes defective for just one of three reasons. Each *can* be repairable, but the time and expense involved generally means it's better to purchase a replacement. The three primary reasons a motor stops working are binding, an open, or a short.

Binding is a mechanical problem. It is caused by any of several things. Lack of lubricant is a common problem, and it is not always easy to cure since many motors are sealed units. The bearings or bushings could be worn out, or they could be lacking lubricant, or both. It's also possible that physical damage has occurred. The shaft might be bent, or the pulley could be misaligned on the shaft.

Binding and other physical problems are detected by examination. There will probably be resistance when you try to turn the motor shaft, but it shouldn't be immovable, nor should it be uneven in its motion. (Also check to see if the shaft wiggles. This is a sign of worn bearings, even if the motor is not binding.)

Open circuits are most commonly brought about by a broken or burned wire or winding inside the motor. In some cases, this break might have occurred at or near the motor's terminal, making it easy to spot. When a wire is broken, current cannot flow and the motor stops. The solution for a break in the wire outside the terminal is to simply repair or replace the wires, or solder them back into place.

If the open (break) is in the windings, it is unlikely that you will be able to do much to correct the problem. However, if the motor is otherwise a delegate for the trash barrel, it might be worth the effort.

Gently remove the insulating covering of the motor windings to ex-

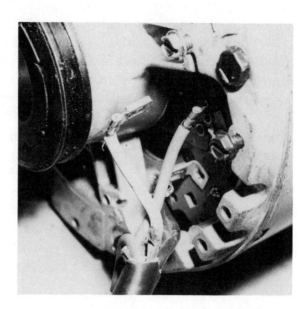

FIG. 5-8 Start your search with the easy and the obvious. Are the wires securely connected to the motor's terminals?

pose the coils of wire wrapped around a core. Unsolder the connection from the wire end on the outside of the winding and slowly begin unwinding it. Sometimes the break is on the first or second layer. Splice the connection in this manner:

1. With a knife or wire strippers, peel off the layer of lacquered insulation surrounding the ends of the broken winding.
2. Twist the two cleaned and pre-tinned ends together, and resolder.
3. Cover the soldered splice carefully with insulating tape, to prevent the possible shorting out of some of the turns of the winding.
4. Rewind the wire, in as close to the original position as possible, and resolder the end to the connecting terminal to the voltage source.

Warning: This process could easily ruin the motor. Do *not* try it on a motor that has a trade-in value.

The third cause of motor malfunction is a short circuit. Somewhere inside the motor windings, the insulation has been scraped or burned off and a few or all of the winding turns have been shorted out. This means that the motor may try to turn, but it will either rotate slowly or will just sit in one position and hum or vibrate.

Most of the time, such a short circuit in the internal winding will lead to an arcing of the voltage from one exposed hot point to another, and the heat will burn the winding open. The cure, unfortunately, is usually to replace the motor.

As detailed above, a motor that has failed due to binding or other physical problems can be diagnosed by examination. The second two causes (opens and shorts) can be detected by some simple tests.

PROBLEM DIAGNOSIS

As a first cut in diagnosing an appliance fault, you should understand exactly in what manner the appliance is failing. For instance, if the appliance has a plug, is it plugged in? This may sound terribly elementary, but it is an essential step in disciplined problem diagnosis. An example is the disposal found in the kitchen sink. Generally speaking, a disposal is often wired as an integral part of the home wiring system, but in some homes there is a little receptacle buried in the dark recesses under the sink, into which the disposal is connected. Putting items under the sink can some times crowd the space and knock the plug out of its ac receptacle.

If a motor has been under heavy load or running for a long time, it will sometimes just stop for no apparent reason. In a situation such as this, look for the thermal overload circuit breaker which, if present on the motor, is usually a little button on the end of the motor case, often colored red. This button should be pushed in, which resets the circuit breaker allowing the motor to run again.

These little heat protection circuit breakers can be found on a variety of motors, usually associated with heavier-duty appliances or electric-powered tools, but they can be found on other appliances around your home.

The motor might directly drive the appliance. It might also drive it through the use of gears and/or belts. Visually inspect all of these parts for proper function.

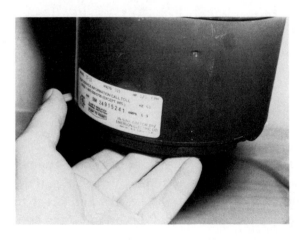

FIG. 5-9 Some motors come with a built-in circuit breaker.

Check all of the obvious things first. More than one person has re-placed an expensive motor, only to find that the actual problem is else-where, and something very simple.

Pay attention to all of your senses. A burned winding will create ei-ther an open or a short, depending on where it is. Either way, the motor won't work. That burning will probably be visible if you can get inside the motor. Whether you can get inside or not, the burned insulation al-most always gives off a sharp odor.

If the appliance is motor-powered, does the motor turn? If the motor doesn't turn, does it make a humming sound or other noise indicating that power is getting to the appliance? Is it getting too hot? (If the motor is getting overly warm, this is an indication of it driving a load that is too heavy, a lack of lubrication, or both.)

TESTING THE MOTOR

Once disconnected from a circuit, with any pulleys, drive belts, or drive shafts removed or unscrewed, the motor can be checked as an indepen-dent electrical device.

Check to see if the shaft is solidly in place. If it rotates freely but wig-gles quite a bit from side to side, the bearings are worn out, and the only cure is to replace those bearings. Lubrication will not help. The only cure is to replace the bearings, or replace the entire motor.

Next, see if you can physically rotate the motor shaft with your fin-gers. If it is bound up, you may have frozen bearings, or overheating may have caused the metal to expand so much the shaft has bound some-where inside the casing.

Especially with permanently sealed motors, once it is frozen in this manner it is useless. If it has no trade-in value, you might as well take the chance of ruining the case and motor to see if you *can* get it operable again. The best way is to take a very small electric drill, smaller than ⅛ inch, and drill a hole through the casing by the bearings or other spots that might need lubrication. (Be VERY careful with this. Before proceed-ing, you should already have the motor apart so that you know exactly where to drill.)

Use a vise or clamp to hold the motor in place. Drill in an upward direction so any filings from the operation will fall outside the motor, in-stead of inside the casing and in the windings or bearings.

Once you have a hole in the casing, you can either use some type of high quality penetrating machine oil, or a silicon lubricant can be squeezed through the opening you've created. Allow the lubricant to soak for a few hours, then again try to turn the shaft by hand. Make sure you allow enough oil or silicon to settle in place so that the shaft spins freely and easily, and listen closely to see that there is no grinding or scraping noise from inside as the motor turns. If the motor is still noisy, try to lubricate it again. But keep in mind that over-lubrication can be almost as bad for a motor (and sometimes worse) than under-lubrication. Both can destroy the motor.

If the shaft stays in place and spins freely, the lubrication may have been all that was needed.

Check to see what type of voltage is required to operate the motor. An ac motor generally runs on 117-vac and can be tested by running test wires with alligator clip endings from the motor terminals to the voltage connectors on the appliance. When this has been done, carefully plug the appliance into a wall outlet and turn on the on/off switch. This way you can see if the motor is now operating correctly. If so, you can reassemble the appliance to see if your lubrication work was enough to get the unit working again.

If you have a dc motor, remember that without a load it can run for only a few seconds without going into high speed and self-destructing. *Never* apply ac to a dc motor. Always apply the correct value of voltage from a dc source, and for just enough time to see that you have freed the motor's binding enough for it to run. Then disconnect the motor, the wall plug of the appliance, and reassemble the unit.

TESTING WITH A VOM

A motor winding is nothing more than a coil of wire twisted around a core. The wires come out to a terminal strip or other connector. It's usally a simple matter to visually inspect those connectors to determine which contacts go to which of the internal windings. After you've determined which leads go where, set your VOM to read ohms in the x1 scale.

The two terminals of an individual winding should show resistive continuity. If the VOM indicates infinite resistance when placed aross the terminals of a winding, there is an open somewhere in that winding. (If there is continuity between two different windings, there is a short between those windings.)

The reading for continuity should be very near zero. However, it should not *be* zero. The wire inside the motor should give a small reading. A reading of zero probably means that you are doing something wrong. It could also mean that there is a direct short between the two contacts.

Just as you should get continuity through a single winding (a reading of close to zero ohms), there should be a reading of infinite ohms be-

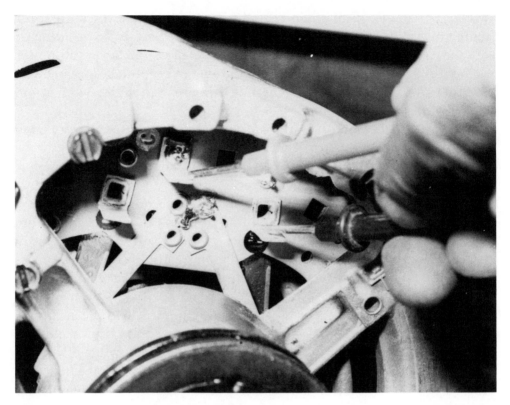

FIG. 5-10 Testing a motor with a VOM.

tween different windings. You should also get a reading of inifinity between any of the leads and the motor casing. A reading of any resistance here (that is, any reading less than infinite ohms) means that the motor windings are shorting, if only slightly, with the case. This is a dangerous situation.

Since this test is both for the motor and for your own safety, carry it out carefully. Set your meter in the highest range for resistance. In the x1 range, a full-scale reading may or may not indicate infinity. In the x1000 or above range, that full scale reading is more likely to mean what it indicates.

Other electrical parts of the motor can be tested with a VOM. If the motor has a capacitor, for example, the VOM can provide a quick, although not entirely reliable, test.

Short out the capacitor first to drain off any charge it might have. If at all possible, disconnect at least one of the two leads. (The order of events is important here: drain the capacitor of charge *before* either disconnecting it or touching it with your fingers.) Now touch the probes of your meter to the capacitor leads. The needle should swing to a low-resistance reading, and then drift back towards infinity. The meter is putting a slight amount of current into the capacitor. As it "fills up," it becomes more and more resistant to accept more. Hence the change in the reading.

MOTOR DISASSEMBLY

The exact steps for disassembly of a motor will depend on the particular design of that motor. The method used by the manufacturer to assemble the unit must be followed precisely, right down to which screws and spacers go where, and in which sequence.

Go very slowly. Take thorough notes. Make sketches where appropriate. Without notes and sketches, you could get inside, fix the problem, and then be unable to successfully reassemble the motor again.

Keep in mind that many of the motors used today are not meant to be opened. This design is meant to help keep dust and other contaminants out. It also means that such a motor is often impossible to repair. It's disposable, just like some of the new razors on the market.

REPLACING A MOTOR

There are times when an exact replacement motor is not available. If you can find a motor of the same physical size, with a shaft of the same thickness and length, operating on the same voltage, and that you can remount in the same position as the original, and in the same manner — then it may be possible to substitute this for the original. However, unless you are well versed in motors and motor types, you are better off working with exact replacements.

In replacing a dc motor, correct polarity (plus to plus, ground to ground) is important. Make sure you solder or attach the screw tap connections to the replacement exactly as they were on the original. Again, your notes and sketches are important. Wire up the new motor exactly the same way as the old one you took out.

MAINTENANCE

There are three basic parts to proper motor maintenance. The first, and easiest, is keeping it clean. The second is keeping it properly lubricated. The third is keeping an eye on all related parts.

Dust, dirt, lint, and so forth are deadly enemies of anything mechanical. Allow a motor to become dirty and it will almost certainly begin to have troubles. Thoroughly clean all motors and related parts on a regular basis. How often depends on how and where the motor is used. A visual inspection every few months is a good idea in any case.

It should be quickly apparent if the suspected motor is designed to accept oiling. These motors need lubrication, but it can present problems if handled improperly. All motors and drive gears should be lubricated at least once every six months (more frequently if suggested by the operating manual). But lubricant must be applied *carefully*. Excess oil can not only damage parts, but it will also collect and gather any dust or dirt in the area, aggravating an already difficult problem.

Use light machine oil (3-in-1 is a good type). Be sure to wipe up any excess spills with a soft cotton cloth. It helps to lightly moisten the cloth with denatured alcohol (*not* rubbing alcohol).

If you get careless and accidentally spill some oil, clean it away immediately. In cases where belt drives begin to slip after an oiling, the inner surface of the belt (where it runs across a drive shaft or pulley) can be "dressed" with either soap or beeswax, just as an automobile drive belt can be dressed when it slips. Just hold the bar of soap or beeswax against the inner surface of the belt as it turns. Excess oil will be absorbed and the belt will begin to "grab" better. You can actually hear the belt-driven device increase in its rotating speed.

In such dressing operations, be sure to use common sense safety precautions. Make sure your fingers, hands, and the soap or beeswax won't get caught in any moving mechanism. Apply the dressing cautiously. Also, check to see if the appliance comes with some type of belt tightening adjustment screws or springs. These may be repositioned to tighten a belt after prolonged use.

If a drive belt still slips after dressing and tightening attempts, it has stretched out of shape and must be replaced. An exact replacement is necessary. Any substitute is only a temporary solution to make an appliance work again. It might *seem* to work okay, but the improper replacement can put a severe torque strain on the drive motor, leading to permanent damage to the motor and to other parts.

This is especially critical in heavy-duty appliances such as vacuum cleaners, electric lawn mowers, and power tools. Replacement motors are a heck of a lot more expensive than replacement drive belts. Don't try to save pennies by using a substitute belt replacement—which will quickly burn out the motor after *seeming* to effect a satisfactory "repair."

Motors driven by ac power often have condensers (capacitors) directly mounted somewhere on the motor casing. These are usually of the *electrolytic* type. This means the condensor is manufactured with an electrical insulating paste between the metal plates of the device. After two or three years time, the paste begins to dry out and the capacitor begins to leak (allow the passage of dc voltage). A capacitor is designed to *pass* ac voltages and *block* dc. As normal maintenance, electrolytic condensers should be replaced after they are three or four years old, even if they still seem to check okay on a VOM test. This replacement of an inexpensive part can save big costs if the capacitor goes completely bad (shorts out) and destroys the motor or the unit's power supply.

Chapter 6
Heating Elements and Thermostats

Irons, heating pads, hair dryers, electric skillets and cookers, electric blankets, toasters, hot plates, coffee makers, and all types of ovens or broiling units make use of some form of heating element. And this is only a partial list. Basically, any appliance that has heat as a function has a heating element to provide it.

Most of the units with heaters are thermostatically controlled, meaning that a device causes an electric switch to turn on or off when certain operating temperatures are reached. Many also have some form of temperature control, usually via something called a rheostat.

The purpose of this chapter is (1) to demonstrate the way these related units work and (2) to show you how faulty devices may be checked and then repaired or replaced.

HOW IT ALL WORKS, BASICALLY

Rub your hands together quickly. The friction will create heat. The more friction there is, the more heat there will be. The less friction, the less heat.

The same happens in an electrical conductor. A wire with low resistance, such as those in the walls of your home, allows electricity to pass through without a large heating effect. Any electrical conductor that offers a high resistance to the passage of electricity becomes a heating element. The energy required to force the current through the resistive wire is passed off into the air around the material as heat.

Heating coils and elements are made of metal alloys of chromium,

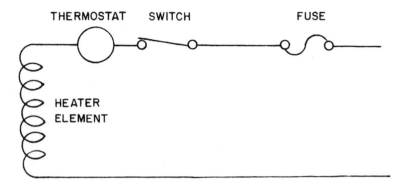

FIG. 6-1 A heating element schematic.

nickel, tungsten, aluminum, and other metals in combinations that are physically strong and durable as well as heat generating. Nichrome, an alloy of nickel and chromium, is one of the most common alloys used for heating elements.

Various control devices, such as thermostats and rheostats, can be placed in line with the heating element. A thermostat is an automatic switch. When the current flow causes the heating element to reach a certain temperature, the thermostat causes the current to be cut off. With an adjustable thermostat, you can set the thermostat to shut off current flow at whatever temperature you select.

A rheostat can also be put in line to govern temperature. The rheostat is a variable resistor that controls the amount of current flowing through the heating element. The more current, the more heat; the less current, the less heat.

However simple or complex the overall system, there are three basic ways in which heating elements and related parts are internally connected to the unit's voltage supply wiring: plug-in, lug connections, and welded or soldered.

PLUG-IN UNITS

Plug-in units have either prongs similar to those on an electrical cord or specially sized prongs or sleeves shaped to slide over the contact mounts on the appliance. This is the easiest type of heating element to disassemble, test, or replace since removal and reinsertion is a simple matter of unplugging and plugging.

There can be two problems, however. If an appliance has been in use for a long period, the contacts may be almost frozen into place by corrosion, by being slightly bent, or by a build-up of grime that has

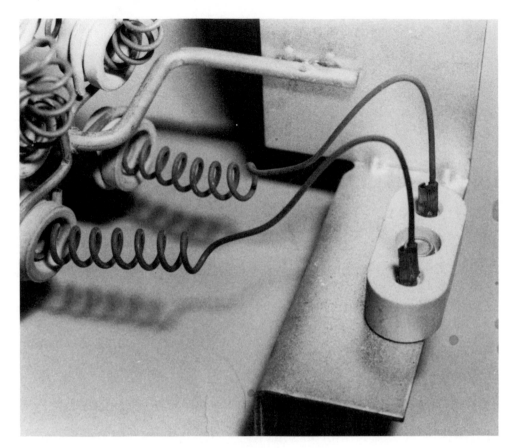

FIG. 6-2 A plug-in heater element.

packed the contact slots. This is especially common when there is water around the heater element and/or the contacts. You'll have to be careful when removing the plug on such a unit, especially since corrosion might have weakened the prongs.

Don't just assume that the element won't come out because of the build-up. Before attempting to use force to jiggle or pry tight plug-in units from their sockets, study the device carefully to see if the manufacturer has placed small holding or set screws somewhere as a locking device.

LUG CONNECTIONS

Sometimes the wires leading to and from the thermostat or heating element are equipped with spade or eyehole lugs, designed to slip over or under contact lugs, screws, or bolts. That's a complicated way to say that the element plugs in, but with wires and connectors instead of more directly.

FIG. 6-3 A heater element with lug connectors.

Before running out to buy a replacement element of this type, be sure to inspect the wires and connectors, and then use a VOM to test that the lugs are properly soldered or clamped onto the wire endings, and are making proper contact. Once again, the biggest enemies are grime and corrosion. What appears to be a fault with the heating element could be nothing more than a lug and wire that is no longer making contact. Often an open or partial break in the lug contact or in the lead-in wire itself can be at fault, rather than a damaged or open element.

Wire and lug troubles can be easily repaired at home, avoiding the cost of purchasing a replacement element. You might have to replace the lug if it is too far gone, but the 5¢ cost of a replacement lug is better than spending $15 for a new heater element.

WELDED OR SOLDERED UNITS

Some elements and their connecting wires are spot welded or silver soldered onto the proper voltage contacts within the appliance. This is a particularly common method of attaching the element wire to the overall heating element unit. Replacing such elements and other spot-welded parts can be difficult, and in some cases will be impossible without the proper equipment.

If regular electrical solder is used to install a replacement, the heat

FIG. 6-4 Spot-welding on a heater element.

generated in making the element work will be sufficient to melt the solder. To prevent the solder from melting, manufacturers generally either spot weld or silver solder any direct connectors to the coil.

Spot welding is just what the name implies. It is a weld (a joining of metals at very high temperature) in a small spot. The result looks like a small indent or burned spot. This creates an almost permanent joint, and one that won't melt or let go unless the temperature climbs far above the operational level, in which case the element wire will melt long before the weld does.

Silver soldering is a special technique for making heat-resistant electrical contacts almost as strong as spot welding. A solder containing silver is used, which melts only at a very high temperature, and once it is in place, the temperature needed to melt it again is even higher. Once again, before the solder itself melts, the element wire will melt.

To replace spot-welded elements, it is sometimes necessary to use special crimping tools, or heat-generating crimp/clamps. Even if you have the proper equipment, spot welding or silver soldering can be difficult. It depends on the circumstances and where the weld or solder is.

WHAT GOES WRONG?

There's not much to go wrong with the heating element. It's one of those situations where it either works or it doesn't. Diagnosis and testing of the heating element and the appliance as a whole relies on the process of elimination.

Basically an appliance that uses heat consists of a power cord, a switch, a pilot lamp, a thermostat, a rheostat, and the heating element itself. (Not all heating appliances will have all these parts. Some will have other parts and control circuits, which can be diagnosed separately.)

Chilton's Guide to Small Appliance Repair/Maintenance
HEATING ELEMENTS AND THERMOSTATS

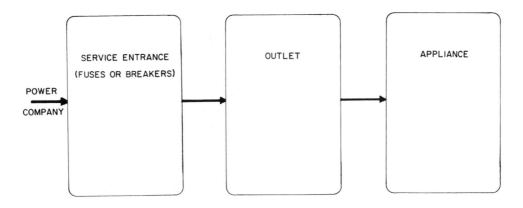

FIG. 6-5 Malfunctions originate in one of three locales: between the service entrance and the outlet; between the outlet and the appliance; or inside the appliance.

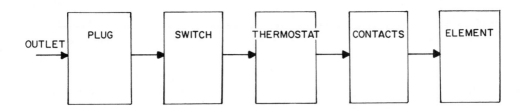

FIG. 6-6 If the problem is inside the appliance, it can only be in a few places: the switch, the thermostat (or rheostat), the contacts, or the heating element.

Keep this basic plan in mind when you are troubleshooting a malfunction.

For a totally nonfunctioning appliance, or one that is blowing fuses, there are two possibilities. The fault could be between the service entrance and the outlet, or it could be from the outlet and into the appliance. This sounds (and is) so simple that many people ignore the obvious, and will tear apart the appliance, then spend the money to replace the heater element, only to find that the whole problem in the first place was nothing more than a circuit overload.

Particularly with appliances that heat, this first step in the process of elimination cannot be ignored. Due to the heavy current draw of anything that purposely generates heat from electricity, problems of circuit overload are more common. (You can usually skip this step,if the appliance has a function other than heating and those other functions work, or if other appliances operate perfectly on that outlet or circuit.)

Once you know for sure that the problem is not in the circuit or outlet, you can begin eliminating the other possibilities. If it's not the cord going to the appliance, then it's in the appliance. If the heating element tests okay, then the problem is inside the appliance between the heating element and where power enters the appliance. And so on.

Much of the time the source of the problem can be spotted visually. The heating appliance draws fairly heavy current, and this means that burns, melts, and other damage can occur. For example, if the wire of the heating element is exposed to view, you might be able to quickly spot a melt in the wire. Or you might see a burned contact or other component that has been charred or damaged from heat. (However, wire elements in appliances that create a lot of heat will become discolored and be perfectly fine.)

Probably the most common failure is a burn-out of the element itself. The heater element is made of high-resistance material and is designed to heat up. However, no matter how well made, each time the element heats, stress is placed on the material. Sooner or later, even the best heater element will give out. How it does so, and how soon, depends on the appliance. The element wire itself might melt. Or sufficient corrosion might develop along the element or at the connectors that the current flow needed to create heat decreases or ceases altogether.

The thermostat might freeze up, either on or off. The rheostat might become so corroded that the corrosion forms a layer of insulation between the internal contacts. Even more common, the contacts between the parts, which are often exposed, will corrode and then either break or simply stop passing current.

Fortunately, almost all of these things are easy to diagnose. Repair usually requires a replacement of that part, but that is almost always far less expensive than replacing the appliance itself.

TESTING THE HEATING ELEMENT

The heating coils and their contacts are often visible when an appliance has been taken apart. If the element is damaged from overheating or age, this can often be seen in the form of an obviously melted wire, a crumbling wire and/or insulation, or other such damage.

If you don't see damage, or if the element itself is not visible, make a continuity check with your VOM. You should get a very low resistance reading (although not one of zero ohms) from one lead of the heating element to the second lead. If you get a reading of zero, you're probably not using the VOM correctly. A reading of infinity means that the wire of the element has broken, has melted, or that corrosion has reached a point either in the element or in the contacts so that current flow is seriously impeded.

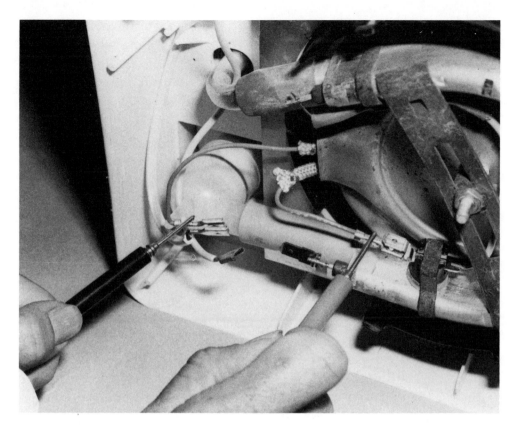

FIG. 6-7   Testing a heater element with a VOM.

Most coils resist ac current. This form of electrical resistance, called *impedance*, cannot be measured with the VOM. But if the meter shows continuity from one lead to another, the element is not open, and is probably okay. Check elsewhere in the appliance for that particular problem.

Many heaters and heavier-duty appliances have more than one heating element. In these cases, separate heating coils are used for various temperature ranges, with switching (either manual or via relays) to shift between the various ranges. This is especially true in the case of heavy-duty electric irons and baseboard or wall space heaters.

Other appliances may have two completely separate elements. An automatic coffee maker is a good example, with one heating element used to heat the water, and a second element used to keep the coffee warm.

If the malfunctioning appliance has two elements and both go out simultaneously, it's rather unlikely that both would have burned out at the same instant. Instead of wasting time testing both elements, look elsewhere first. Both elements are supplied from the same outlet. Except for large appliances that require 220 volts, both are probably supplied from the same wires and the same switch. Start your testing with these common items. (Chapter 2 gives you all the information you need on testing cords, switches, and so on with your VOM.)

Before putting away your meter, also test for continuity between both element contacts and ground. You should get no reading (or, rather, a reading of infinite ohms). This shows that the element is isolated from the metal chassis. A reading means that something has shorted out. It could be something as simple as moisture, or something as deadly as a wire touching the metal chassis of the appliance. Either way, if you get a reading between the heater element and ground, *do not* attempt to use the appliance until the source of the problem has been found and repaired.

A problem that is a little more difficult to track down is an appliance that works, but only partially. It might heat up when it is turned on, but not enough heat energy develops to do the job properly. In these instances, only one of the several heating coils or elements may be faulty. Even more likely, the element is simply wearing out. This will show by a high, but not infinite, resistance across the element (as shown by testing with a VOM). Pinpoint and repair or replace the bad element, and the device is at like-new efficiency once more.

REPAIRING THE ELEMENT

When a heater element is malfunctioning, the usual repair is to replace that element. Sometimes this is simply a matter of unplugging the old element and plugging in the new one. Other times it might require removing the element wire and then carefully fishing the replacement wire through the insulators.

Some very simple heating coils, such as may be found in inexpensive room heaters or in small one- or two-burner electric hot plates, may burn open from aging or through being used for too long a time in the high-temperature thermostat range. In emergencies, or when you expect to use the device mostly in the low or mid-range heating mode, a quick repair can be made by slightly stretching the coil so the two open ends slightly overlap and touch each other. Special clamps can be used to hold the two ends together. Such a "repair" will eliminate one or two turns of the coil, so its overall resistance (impedance) will be slightly lower (resistance is determined in part by the total length of a conducting material). The patched coil will not get as hot as a new replacement, but on low or medium heat it may last just as long as the original did on high heat.

An inadvisable method of repair is to pseudo-spot-weld the wires. This involves bringing the two wire ends close together, so that they just barely touch. Carefully apply electrical current, watching the spliced coil. A flow of current will jump (arc) from one broken wire to the end of the other, and this quick arc will spot weld the splice.

THERMOSTATS

The "brain" of any heating device is the automatic control that varies the temperature of the appliance according either to diurnal (day to day) conditions or the need of the operator. This is the thermostat. Several types are in use in any modern household, from the basic heating-cooling thermostat used to control the home's central heating and cooling units, to the small internal devices used to automatically maintain the proper temperatures in hot water heaters, coffee makers, or baking ovens.

There are several types of thermostats, but most work on the basic principle that metals of two different types will expand or contract unequally when heated. Two metals in what is called a *bimetal strip* will bend or change shape according to temperature changes, and this movement is used to operate an automatic electrical contact switch. The opening and closing of this switch turns the heating (or cooling) element on and off.

When bimetallic thermostats are controlled by a rheostat, the turn-

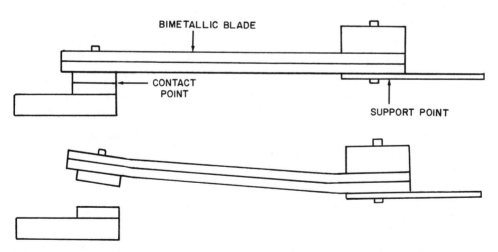

FIG. 6-8 The function of a bimetallic thermostat, closed (*top*) and open.

ing of the knob changes the relative position of the two pieces of metal. It takes more or less air temperature change to make the two metals expand/contract enough to make the electrical contact.

Very simple thermostats are used as the heat controls for water heaters and many coffee makers. In appearance, these seem to be dull silver-colored "lumps" on a short piece of wire. Inside the "lump" is a bimetallic switch. Other times the thermostat will be more exposed; in some the bimetal strip will be quite visible.

A few of the control devices are *thermisters*, or *varistors*, which are semiconductors (transistor-like devices) that are subject to changes in internal resistance when different voltages or amounts of heat are applied. The changing resistance becomes the controlling effect, or "brain" for the appliance.

In total home heating and cooling operations, it is often discovered that a single centrally located thermostat does not do the job of control properly. The spot selected for the thermostat location is too hot or too cool in comparison with the rest of the home. For this reason, many people decide to switch over to multi-stage thermostats, with separate thermostat control used to set temperature for different areas in the home. (Multiple thermostats can be complicated to test, but it can be done by following the process of elimination.)

SOLID-STATE THERMOSTAT CONTROLS

Newer appliances utilize computer-type switching devices. A heat-sensing thermostat can be built right into a tiny IC (integrated circuit) chip, along with transistorized switching networks. As the ambient temperature of the air or the material being heated or cooled changes, the miniature thermostat "reads" the change and "instructs" the switching network to turn the appliance on or off.

These sensing chips are relatively inexpensive and cannot be repaired (although they do go bad). If this is the type of thermostat control on some of your home appliances, you will have to obtain an exact replacement of the chip from the manufacturer or service dealer. And more likely, you will be required to purchase the chip and all of its associated circuitry on its mounting board.

TESTING THE THERMOSTAT

Before wasting a lot of time in testing a suspected thermostat, first visually examine the thermostat and any contacts. The whole problem could be something as simple as dirty or pitted contacts. After you've determined that the thermostat *is* the most likely cause of the trouble, you can go on with testing of the unit.

Since thermostats are switches, and normally open when disconnected, test meters are of limited use in checking to see if the device is faulty. The best test is operational. Once a heating element and other components have been tested okay, and if you know that voltage is being supplied to a device, poor heat control is a sign of a bad thermostat. Repair is usually by replacement of the thermostat unit as a whole. Replacements are usually quite inexpensive—from $3 to $10 in many cases.

Sometimes you can go farther with testing. Other times you cannot. It depends on how the thermostat functions in that particular appliance.

First place a wire across the two prongs of the plug to the appliance (after it has been unplugged from the outlet, of course). This shorts out the thermostat and the appliance in general.

Set your meter to read resistance in the x1 range. Now touch the two leads of the thermostat. With the on/off switch of the appliance in the on position, you should get a reading of continuity across the thermostat.

FIG. 6-9   Testing a thermostat with a VOM.

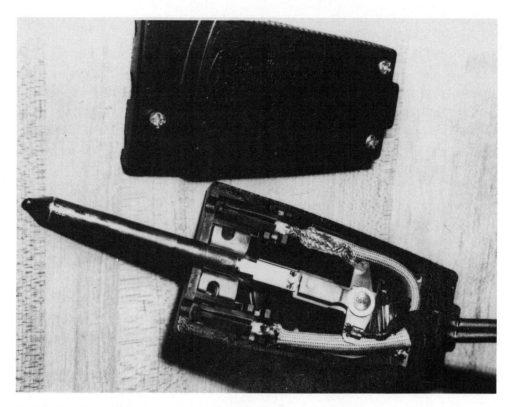

FIG. 6-10 Thermostat control assembly for an electric skillet.

This shows that the contacts of the thermostat are closed, and that current should be flowing.

However, this test doesn't tell you if the thermostat is fully functional—only that it should be allowing current to flow when the appliance is first turned on. If the contacts are fused together, the thermostat will not be able to do its job of shutting off the current. If the thermostat breaks contact too soon, due to maladjustment or wear, the heating element won't get enough current.

Sometimes a meter may be used more successfully for thermostat checking in some appliances, such as a hot water heater. Set the VOM in a high ac voltage range (above 250 vac), and place it across the voltage source (input) to the water heater. Rotate the temperature control rheostat from the lowest to highest temperature setting. At some point, you should be able to hear a light "click" as the thermostat contact is made. At this point, the voltmeter should indicate a *drop* in voltage as a load (voltage being applied through the thermostat to the heating element) is applied. If you hear this click and are able to read this voltage drop, but the water heater still does not go on, then the problem is probably in the basic heating element under the tank. Due to the construction of a water heater, it is best to replace the entire water heater when a heating element goes, especially if the unit is more than eight to ten years old.

Similarly, a voltmeter may be used to check thermostat or thermistor operation in a small coffee maker. With the unit disassembled for metering access, measure the ac voltage from one lead of the electric cord across to the *input* side of the thermostat. If you read voltage there, but cannot obtain a voltage reading from the same side of the ac cord to the *output* side of the thermostat device at a time when that control should be turning on a heating element, then the thermostat is likely bad, or the heater element has opened. Repair of either is a matter of replacement.

TIMERS

Although a timing device isn't exactly a thermostat, it can be used to start or stop current flow at preset times. Timers are simple clock mechanisms, with the clock timer on a rotating wheel. Electrical or switching contacts are placed on the moving timer, so that at certain preselected times a heating, cooling, or cooking device will be automatically turned on or off. Many radios, coffee makers, space heaters, and electric ovens have such timers built-in as a basic part of the appliance.

Often, a timer-controlled appliance will have a built-in outlet, much like a wall outlet socket. This means that the unit is designed so the owner may operate a separate device off of that timer, or that the outlet has

FIG. 6-11 A switched outlet is used in some appliances to supply power to another piece of equipment.

been designed as a separate voltage source to *bypass* the timer. Examples of this are video tape recorders that may be plugged into a wall outlet. The video machine operates off of an internal timer, preset by the owner. The unit also contains an outlet plug so the TV set can be connected directly to the VCR, but the TV set is *not* on the timer. This outlet serves as nothing more than an extension cord or extension plug for the owner, saving the wall outlet for use by some other appliance such as a lamp. Other controlling switches may be located in conjunction with thermostats. These may be *humidistat* controls to adjust operation according to humidity, or they may be some form of electronic air-cleaning or air-filtering controls. All are switches of some type, and they may be checked and repaired or replaced.

Occasionally, in less-expensive small appliances, temperature is varied either by a switching arrangement or by a variable resistor (rheostat). In switch set-ups, different positions are used to either change the amount of current fed to a particular heating element, or are used to switch the voltage/current feed from one heating element to another.

In rheostat arrangements, a relatively high-wattage (high power dissipation) variable resistor is used as a *voltage divider*. Changing the rotary position of the variable resistor raises or lowers the voltage pressure applied to the heating element. Less voltage means less power will be dissipated as heat energy when current passes through the heating element. Changing the rheostat to increase the voltage will result in a heavier power dissipation through the element and more heat energy.

MAINTENANCE

Because there are so many switches and moving or slide contacts, poor operation of a heating element can be a simple matter of lousy electrical contact. There are several methods for burnishing, or improving the continuity of a movable electrical contact. Both surfaces may be rubbed lightly with the rubber eraser at the end of a pencil. Hardware stores carry small contact burnishing tools, a type of small file or sanding device that can be used to scrape away grime and oxidation from contact points.

An aerosol spray contact cleaner provides another possible cure. A warning, however. Be *sure* to purchase *nonlubricating* contact cleaner. Several aerosols are on the market that are specifically designed for use on television tuners. They contain a lubricant, or a silicone concentrate, along with the contact cleaner (alcohol or some other fast-evaporating cleansing agent). The lubricated sprays are fine for TV tuners, but if they are used on switch or rheostat contact points in an appliance designed to heat, the lubricant will remain as a dust-gathering contaminant. A build-up of dirt will quickly cause the switch, contact, or rheostat to go bad again, possibly damage it severely.

Most heating elements have no moving parts unless a fan driven by an electric motor is used to disperse the heat. Maintenance is a matter of simple cleaning and keeping dust, oil and grime off the element. Generally no more than once a year, use the reverse blowing setup on a vacuum cleaner or an air compressor to blow dust off these elements, or wipe gently with a dry cloth (moisture is a conductor, and if the coils are wet when electricity is applied, a short may develop and destroy the element).

Simple cleaning of heater elements is a procedure many people tend to forget. Check to determine if the appliance you're working on has an air filter. Filters in forced air systems clean the air that enters the appliance. While this protects the inner components, those filters can cause a problem. If clogged, the internal temperature can rise, causing the thermostat and other controls to operate falsely. The result is improper operation and possible damage to the appliance.

Filters of the throwaway type need replacement from time to time (check your owner's manual for frequency of replacement). Other types of filters may be removed and cleaned with cool water.

Grills or registers and thermostat devices themselves also require frequent cleaning, since dust and small objects can easily fall into them. Cleaning can usually be accomplished by using a vacuum cleaner or an air blower.

In any heating or cooling device, keep the fan blades clean. When they become loaded with dust and lint, as can often occur if left unattended, their ability to move air is cut down. Most blowers/fans have blades that are easily accessible for cleaning. You can reach them with a small brush or soft cloth. If the blades are not so easily reached, you should disassemble, clean, and adjust the blower/fan at least once a year.

Any belt that powers a fan blade should not be stiff or too loose as it winds around any drive pulley. Properly adjusted, a drive belt should have from ½ to ¾ inch of slack, or play. If the belt appears worn, replace it. A maladjusted drive belt will not only cause a mechanical malfunction, but it could also cause the thermostat, other controls, and the heating element itself, to malfunction.

Thermostats sense the ambient temperature and then turn its associated appliance either on or off accordingly. A coating of dust or grime will make it unable to detect any but the widest temperature changes, and the thermostat will lose efficiency. Clean them annually. This holds true for the tiny thermostat controls for coffee makers, electric skillets, and other cooking/heating home appliances.

Chapter 7

Appliance Purchase and Care

Most modern appliances require little or no maintenance. With some, the disassembly required for internal maintenance isn't worth the risks. Small appliances in particular are often meant to stay closed.

However, there are some things that you *can* do to lengthen the life of your appliance, while reducing the number of times that you have to call in the serviceperson or replace the appliance.

Most of it is common sense, and proper cleanliness. It doesn't require a lot of time or effort. And for that slight investment, you can increase the life of your appliances while cutting the costs to keep your appliances running perfectly. There is no better way to repair an appliance than to keep it from breaking down in the first place.

Start thinking of preventive maintenance even as you are making the purchase. Will the appliance do the job you want it to do in your home? Does it have the features you need? Will it fit? Is it compatible with the wiring in your home? This involves some careful shopping on your part, and even a few measurements and calculations.

WILL IT DO THE JOB?

It's common for a new purchaser to expect an appliance to do something it can't. This both causes disappointment for the owner, and could tempt the owner to use the appliance in a situation where it could be damaged.

This is the first problem. The appliance you buy should be capable of doing what you want it to do. And you shouldn't try to get it to do things it isn't designed to handle.

Putting undue strain on the small appliance will cause something to give, and what will "give" is the appliance.

The more complex the appliance, the more likely it is that something will go wrong sooner or later. So don't necessarily buy the most feature-filled appliance you can find. If you won't need those features, they are just more things to cause you trouble.

PHYSICAL SIZE

Physical installation is not an issue with most small appliances, but some can present a problem, both in where they will be placed while in use, and where they will be stored while not in use.

That kitchen counter may not have enough room for both the toaster and the coffee maker without having one of them dangerously close to the stove. That will mean that you'll either have to find another counter for the appliance, or find a place to store it while it is not in use.

Many of the small appliances used in and around the home are used for a short time, then put away until they are needed again. Store it properly. Improper storage can spell the ruin of that appliance. (Your electric drill, for example—do you know it's not rusting away in a damp basement?)

POWER REQUIREMENTS

Be sure you know the power requirements for the appliance you intend to purchase. In particular, appliances that heat should be checked for the amount of current that they draw. The larger the appliance, the more important this is. In some cases you'll have to provide a separate outlet just for that appliance. Power consumption should be listed clearly on the appliance label. If it's not, ask the dealer to show you where the rating is listed in the instruction manual.

The amount of current can be roughly estimated by using the formula $I = P/E$, where I is the current in amps, P is the number of watts consumed by the appliance or other device, and E is the home voltage. (This last will be either 117 or 234.) To use this forumla you'll still need to know the number of watts consumed.

For example, the current used by a 60-watt light bulb is 60 watts/117 volts, or .51 amps (approximately ½ amp.) A device that uses up 1400 watts will draw 1400/117, or about 12 amps.

Be sure not to come too close to the 15-amp rating of the standard circuit. If you plug that 1400 watt device into a circuit, don't think that you can safely load up that same circuit with the 355 watts that brings the circuit to exactly 15 amps of current draw. If the line voltage drops temporarily to 110 vac, the current flow with that 1400-watt appliance

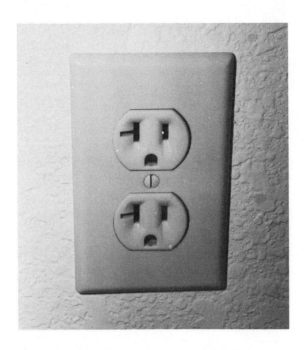

FIG. 7-1 A dedicated outlet. The notch in the outlet indicates that the outlet is for 20-amp service.

will increase to 12.7 amps. Everything else on that circuit will also draw more current. So if the circuit is presently carrying 3 amps (such as six lights), and you plug the new 1400-watt toaster into the same circuit, you're already at 15 amps total. If the line voltage drops, you'll probably blow a fuse or breaker.

Think carefully about where you will be placing the new appliance, especially in regards to power consumption. If the spot you plan to install the appliance has power available only through a circuit that is already at capacity, the new appliance could easily push that circuit beyond its ability to supply power.

BEFORE YOU BUY

"You get what you pay for" is true in many respects. A difference in price of $10 on a $500 appliance doesn't mean much. But if one appliance has a $500 price tag and a similar one is selling for $300, you can bet that there is a reason. (This rule does not apply to items that are on sale.)

Look for good physical construction, whatever the appliance is. Panels should fit well and be fairly solid. Switches should have a good feel to them, and not be sloppy or loose.

Ask the dealer for a demonstration of how you would go about routine cleaning. Don't believe him if he says, "None is ever needed." That appliance might require very little cleaning or other maintenance steps, but all will require some—certainly more than just wiping down the outside case.

Ease of servicing is also important. This shows a good design, and a manufacturer who has bothered to think things out before putting that product on the market. Ask the dealer about servicing and parts availability. He should have at least some idea of where you can get parts locally if you need them.

A few manufacturers have taken the time to design their appliances intelligently. Parts that tend to wear out, such as heating elements, are easily accessible. An unfortunately large number of manufacturers seem purposely to design their products so that you have to ruin the appliance to get inside. For example, on a particular hair dryer, the main case is held together by screws, but before the case comes apart you have to remove the front nozzle-like piece—and this is glued in place. The only way to get it off is to break it off.

Visually examine the appliance before you buy it. You should be able to see how it comes apart—or how it has been designed to *not* come apart.

BRAND SHOPPING

The single best way to know which brand and model to get is to have a personal recommendation from a friend who owns one. If it has worked flawlessly for years for your friends, it's a fair bet that the design and construction are sound.

Some people have a brand name preference; others prefer to save a bit and buy a lesser-known brand that comes with a good reputation or recommendation.

You *do* pay extra for the name, but not as much as you might think. There is usually a reason that the well-known brand costs more. Part of the reason is simply that they can charge more due to their reputation. More important to you is that the extra cost pays for a number of services.

The manufacturer might have large research facilities, used to find and correct faults in previous designs. Others have special service schools for their field representatives. A few even have toll-free numbers for their customers to use. More important, and also as a general rule, the major brand names will make it easier to find parts, and to find service when it is needed.

THE DEALER AND THE WARRANTY

When you buy a new appliance, you are also buying a warranty that says that the appliance will work as specified for a certain period of time. If it doesn't, it will be repaired free of charge.

You can run into several problems here. First, if an appliance is still under warranty, *do not* attempt to repair it or even open the case. This will probably void the warranty. It can be expensive to find out.

Before you purchase the new appliance, ask what the warranty period is. The longer this period, the better. In most cases it will be between 30 and 90 days.

Just as important as the warranty is to know who does the servicing. If the dealer doesn't, he should at least be able to tell you who does. If he has no idea who can handle the servicing—and particularly servicing while under warranty—you might start considering going to another dealer.

The Yellow Pages in your area will give a listing of companies that offer repairs for large appliances. Not many of them care to bother with small appliances. This can be a problem, especially if you're trying to track down a reliable company on your own. The dealer *should* be able to help.

It is an unfortunately common practice these days for the dealer to have no servicing available. One of the worst of the new policies you see is the tag on the new appliance, "Do not return this appliance to the dealer. Send it directly to the manufacturer." But that manufacturer might be on the far side of the country, or even outside the country. Shipping costs can get expensive very quickly. And if you don't properly package the appliance, it can get damaged in transit.

Almost as bad is the amount of time involved. Your coffee maker has failed. So, you get the packaging materials ($$$) then drive to the post office ($$$) to ship it back to the manufacturer ($$$). It will take anything from a few days to more than a week before it will arrive.

With all too many companies, the package will sit unopened for a number of days. It will almost certainly take a week before someone has fixed that appliance, and a month isn't uncommon. Then there's another week before it arrives back in your hands again.

Many appliance owners don't even bother. They'll go out and buy another coffee maker—one from another manufacturer, hopefully.

If the appliance in question is very expensive, or very large, the problem is compounded. Throwing away a $25 coffee maker is one thing. Tossing out a fancy $400 broiler oven is quite another.

Consequently, it is usually better to make your purchase from a dealer who also offers warranty repairs and long-term service. However, just because the dealer has a service bay doesn't mean that he will provide warranty work on your purchase. If that small appliance fails, and it says right on the box, "Do not return to dealer," he is not obligated to provide warranty work.

If you are promised warranty work, be sure to get it in writing. Otherwise it could be just a sales ploy, and you'll find yourself shipping off that appliance to a distant manufacturer anyway.

THE INSTRUCTION MANUAL

All new appliances come with a variety of papers, stickers and tags. It's a good idea never to throw out anything that might be of use in the future, especially any instruction or installation manuals. Make sure that you have one. If you get home and find that something is missing, contact the dealer and the specific salesclerk.

Then read that instruction manual! On more complicated appliances, and large appliances, read the installation instructions, even if you didn't personally handle the installation.

Even if you have a separate identification tag, jot down the model and serial numbers on the instruction manual. Sooner or later you'll want these, and it's good to have all the information in one place. Some manuals will have at least the model number already on it; other manuals cover a variety of models. In this case make note of the specific model number. Also put all receipts with the manual.

You're not done yet. Before you put the manual in the file, be sure that anyone else in your house who will be operating the appliance, or who even *might* be, has also read through the information and understands how to operate that appliance.

In some cases a shop manual will be available. The cost for this manual can be steep. You'll have to decide for yourself whether or not you want to spend that money. At very least, inquire about the availability of such a manual. If you don't want it now, you may in the future.

INSTALLATION

Proper installation may not sound like a maintenance step, but it is. It's also for safety. With many small appliances, installation involves nothing more than plugging it in. If this is all that your particular appliance requires, then other than making sure that the outlet can handle the load, you have no worries about installation and you can skip the rest of this section.

A few small appliances have special installation requirements. You should have little trouble installing any standard home appliance yourself. By handling the installation yourself, you also know exactly how things went in, and you'll know how to *un*install the appliance (and how to reinstall it) if servicing becomes necessary.

If someone else is going to be handling the installation, try to be there to make sure that it's done right. Whether you're there or not, go through the installation instructions and double check the work. (If *you* did the job, triple check it.)

Electrical installation involves everything from the service entrance to your home to the contacts inside the appliance. (Yes, there *are* times when the manufacturer expects you, or the installer, to open the appliance and make the final connections.) As mentioned earlier, the outlet has to be of the right type, and has to be able to safely supply the appliance. Grounding is very critical. Do not try to defeat the safety grounding by doing something silly like clipping off the third prong. If you don't feel competent to provide the correct grounding, spend a little extra to have a professional electrician come out.

The installation instructions, and usually the owner's manual, will tell you if special ventilation is needed, and if so, how much. If the appliance has an internal fan, proper venting could be a critical factor. Don't go for the minimum. It's impossible to have ventilation that is too good, but all too easy to have venting that is inadequate. Inadequate venting spells trouble, the least of which is a shortened lifespan of the appliance.

Many small appliances come with an infuriatingly short cord. The manufacturer's excuse is that this reduces the chance of someone accidentally snagging the cord and knocking the appliance off the counter. What it actually means is that you will probably have to use an extension cord. Using an extension cord should be a last resort. When you must use an extension cord, be sure that it is of the right type and gauge for the appliance. Zip cord meant for lamps cannot safely supply power to an electric fryer. The dangers involved are very real (see Chapter 3).

GENERAL MAINTENANCE

Talk to any service technician and you'll hear again and again the value of preventing problems. A few simple maintenance steps might take ten minutes now. Not doing them can cost you hundreds of dollars later.

The first step is cleaning. How you go about this will depend on the appliance and the circumstances. For a coffee maker this will mean regular cleanings with vinegar or a commercial cleaner designed for that purpose. A hair dryer will require a regular removal of hair and dust so that it doesn't bind up the motor.

Use common sense. As always, your own safety comes first. For example, don't start swabbing down an appliance if it's still plugged in and if the water can get anywhere near a live contact. And don't dunk the appliance in water unless it is meant to be submersed. (Most aren't, and the instruction manual almost always clearly specifies this.)

Think about what it is you're doing before you just lunge ahead. Some parts are delicate and can be damaged by rough handling.

Be sure to use the appropriate cleanser for the job. A harsh, gritty cleanser is unsuitable for cleaning a shiny, porcelain surface. It can also scrub off many of the nonstick surfaces used in modern appliances.

Obviously you won't be taking your appliance apart every couple of weeks just to clean out the dust. Most of your cleaning will be limited to the exposed parts. But occassionally it's worthwhile to do some interior cleaning on many appliances.

The idea is to reduce the amount of time and work. If something has

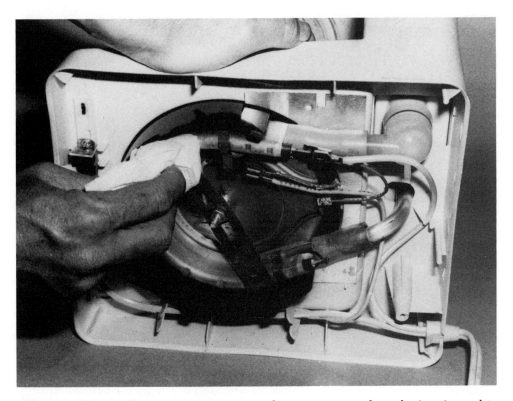

FIG. 7-2 If the appliance is open anyway, take a moment to clean the interior and to check it over for signs of wear that may lead to future trouble.

gone wrong inside and you have to disassemble the appliance anyway, spend a few extra minutes to clean the inside even before you begin working. (It will make working on the appliance easier and safer anyway.)

Any time the appliance needs to be opened, give the insides a checkup. Carefully, with the appliance unplugged, visually and manually check things over. Quite often you'll be able to see something that is starting to go wrong. A switch might show signs of burning; a drive belt might be sagging or show signs of damage; there might be water inside, or excessive corrosion, showing a plumbing problem.

This is a good time to lubricate any motors that can be lubricated. Be very careful to not overlubricate. Usually a drop or two will take care of things. Use a clean rag to wipe away any excess or any spills. Many of the motors used in appliances are sealed units, and generally they require no lubrication.

Whenever you do anything, make notes and sketches for future reference. The notes should include a record of all maintenance done, and anything you've noticed that suggests a possible future problem. For example, if a switch shows a bit of discoloration, make a note of that. Next time you look inside see if it has become any darker (or lighter).

DAMAGE TO THE CASE

Most of the physical damage done to appliances is the result of carelessness. Someone drops a hammer and chips the procelain, or an appliance is carelessly yanked off a counter and crashes to the floor. Your primary concern with any major damage to the case isn't cosmetic, but the effects on the inner workings of the appliance. If that dent causes some moving part inside to rub or bind, don't try to operate the appliance until the damage is repaired.

Some case damage *cannot* be safely repaired. You may be able to buy a replacement part; if you can't, and if the damage is affecting the operation of the appliance, it is probably time to go shopping for a new one.

Some case damage *can* be repaired. Chips in porcelain are an example. Chips in procelain are usually just cosmetic. But they can also lead to rusting of the metal panel beneath and to other problems. Bottles of liquid porcelain are available at most appliance stores and hardware stores. The liquid is simply brushed on. Liquid porcelain is usually made for larger appliances. Many of the smaller appliances that use porcelain finishes are of bright colors not available to the public. If you can't find the appropriate color, you have two choices. One is to find the closest match you can and be satisfied. The second is to contact a custom paint and surfacing store to see if they can come up with something to suit the problem.

There are two keys to using liquid porcelain successfully. The first is in the preparation. The liquid will not adhere properly to a surface that is dirty or wet. Second, take your time, and do your best to have the patch job blend in. Blobbing it on makes a noticeable and unsightly repair.

Better than brushing it on by hand is air brushing. Most people don't have the skill or equipment to do this properly, but it is an alternative. If your own attempts aren't working out very well, consult the Yellow Pages in your area under Appliance Refinishing. It's expensive, but it may be less expensive than buying a new appliance. Explain the situation over the phone, get an estimate, and make your decision from that estimate.

Small dents might be repairable. Sometimes you can simply pop a dent out by pushing from the inside. Other times a dent puller (for automobile bodies) can be used. In many cases dents are not repairable, due to the size and nature of the small appliance. The same applies to breaks in plastic casings or casings made of any other material.

Section Two

Appliances

Blenders, Mixers, and Food Processors

Appliances made to blend, mix, or process foods usually consist of little more than an electric motor, a variety of cutters, mixers, or stirring blades, and a series of controlling switches. However, the wiring to the switches for these units may be quite complex. In any disassembly, make sure to draw diagrams and take notes concerning the proper connecting points for all of the wires inside.

Blenders and food processors use a high-speed motor to drive chopping or mixing blades. Switches (or a rotating rheostat) are used to control the speed. In addition to the normal small-appliance electrical problems involving power cords, plugs, on-off switches, and a faulty electrical motor, these units may also develop problems involving leaks, excessive vibration, jammed parts, bent or broken blades, and bent or broken motor drive shaft.

As always, before disassembly read through the instruction manual, and visually inspect everything for any obvious cause of malfunction.

DISASSEMBLY

To get inside the appliance, first carefully remove and set aside any glass or ceramic parts, such as the blender or mixer bowls or pitchers.

Countertop appliances hold most of their screws and bolts on the bottom plate, although you will occasionally find screws tapped downward near the motor shaft, under the blender bowl location. Be careful before removing any of these, for they may be motor or switch mounting screws instead of disassembly screws. If motor or switch mounting

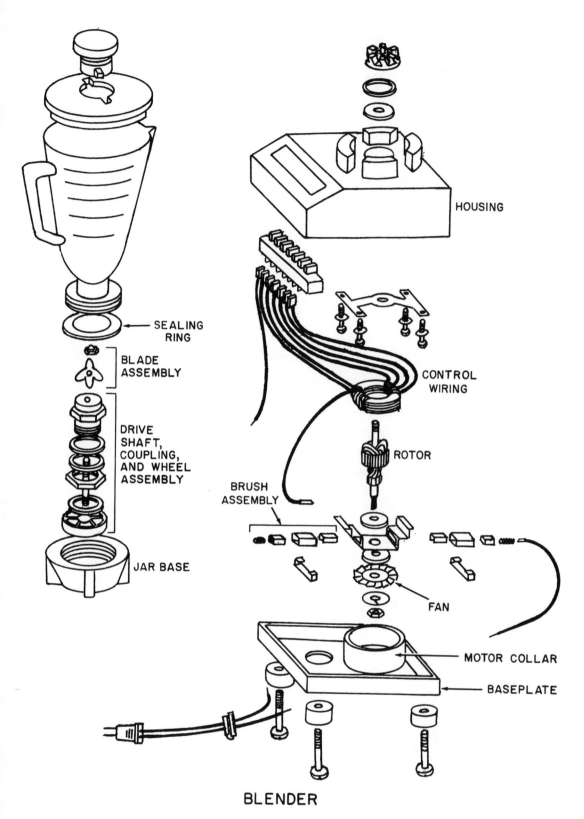

SEALING RING

BLADE ASSEMBLY

DRIVE SHAFT, COUPLING, AND WHEEL ASSEMBLY

BRUSH ASSEMBLY

JAR BASE

HOUSING

CONTROL WIRING

ROTOR

FAN

MOTOR COLLAR

BASEPLATE

BLENDER

Chilton's Guide to Small Appliance Repair/Maintenance
BLENDERS, MIXERS, AND FOOD PROCESSORS

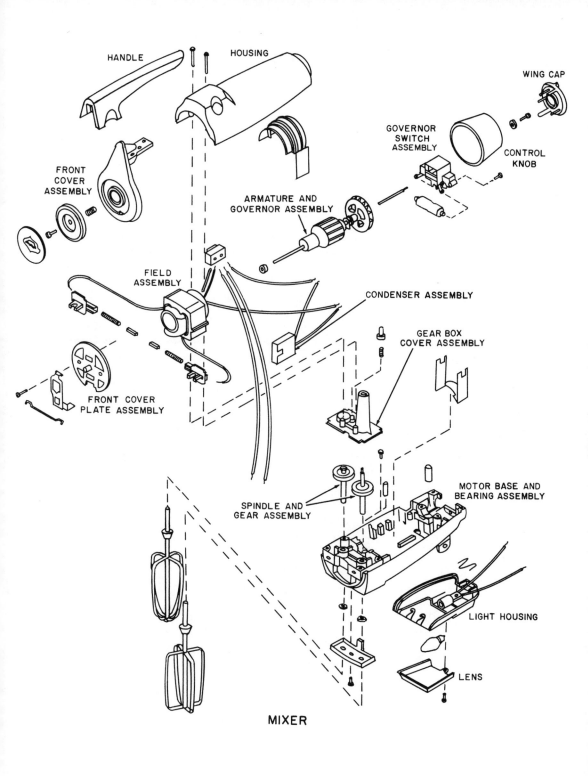

HANDLE

HOUSING

WING CAP

GOVERNOR
SWITCH
ASSEMBLY

CONTROL
KNOB

FRONT
COVER
ASSEMBLY

ARMATURE AND
GOVERNOR ASSEMBLY

FIELD
ASSEMBLY

CONDENSER ASSEMBLY

GEAR BOX
COVER ASSEMBLY

FRONT COVER
PLATE ASSEMBLY

MOTOR BASE AND
BEARING ASSEMBLY

SPINDLE AND
GEAR ASSEMBLY

LIGHT HOUSING

LENS

MIXER

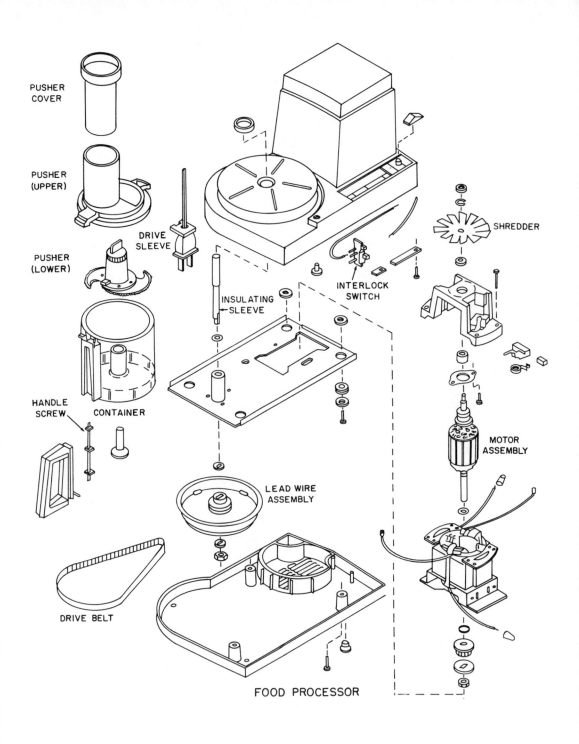

PUSHER
COVER

PUSHER
(UPPER)

DRIVE
SLEEVE

PUSHER
(LOWER)

HANDLE
SCREW

CONTAINER

INSULATING
SLEEVE

INTERLOCK
SWITCH

SHREDDER

MOTOR
ASSEMBLY

LEAD WIRE
ASSEMBLY

DRIVE BELT

FOOD PROCESSOR

screws are disconnected with the bottom plate still on, the motor or switch could drop down inside, causing damage or loss of parts.

Hand mixers are constructed much like electrical drills. In disassembly, the body (case) generally bolts together as two halves. Motor, gears, switches, and drive shafts are located inside the body; the various beating attachments snap into place. Countertop mixers usually have the motor in the base, with gears or a drive shaft extension reaching up to drive the beaters.

Some base plates have snap-on fasteners, which may be popped or lightly pried loose by hand or with a screwdriver. Don't force anything apart until you have made a thorough visual search for hidden screws or clamps.

Study the motor mount arrangement. Speed changes can be effected by changing the position of the motor; in this case the drive shaft has different diameters at the points where it is moved to drive the appliance. Other types of speed controls include sets of gears, or switches in the motor that vary its turning speed.

Placement of bearings, bushings, gears, and drive shafts and the amount of play between parts are critical. Move parts gently by hand in order to learn the mechanical actions of the appliance for proper operation. This will be of help in correcting any jamming or binding difficulty.

If you are taking apart any appliance, review the periodic cleaning procedures that should be part of your standard home appliance maintenance program. Dirt, dust, and spilled food are major appliance enemies. Not enough oil in places where needed, or too much oil where not needed, can become another hazard preventing correct function of the unit. Determine where oiling will help and where regular cleaning can lengthen the service life of the device.

As with all appliances, study any operation or maintenance booklet that may have come with the machine.

Troubleshooting Chart

Possible Causes	Solution

Problem: Machine fails to start, blows fuses or breakers, pilot light fails to light, on/off switch does not work.

Possible Causes	Solution
Wall outlet, power cord, plug, power switch, fuses/breaker, internal reset button.	Check/repair/replace

Problem: Voltage at switch ok, but motor does not energize.

Possible Causes	Solution
Faulty motor	Repair/replace

Problem: Appliance works, but difficult or impossible to change speeds.

Possible Causes	Solution
Speed switches	Test/repair/replace

Problem: Can hear motor turn on but nothing moves.

Possible Causes	Solution
Binding of some part	Disassemble/test
Motor jammed	Test/lubricate/repair/replace

Problem: Appliance seems to be working, but blending operation/function is not satisfactory.

Possible Causes	Solution
Improper blade/cutter used	Check manual for proper blade
Broken/bent blade/cutter	Repair/straighten/replace

Broiler Ovens

Broiler ovens, made for toasting bread, broiling meat, and warming up food, are seldom very complex. There is a heating element, temperature control switches, an on/off switch, the electrical cord and plug, and occasionally a thermostat in the more expensive models. Broilers and toaster ovens are put together in much the same way, and the components used are also much the same. There is an important difference between them, however. A toaster oven cannot be used to broil, and trying to use it to broil is likely to start a fire.)

Another feature often found in these appliances is a heat-, smoke- or time-controlled safety switch that turns off the heating element automatically under certain conditions. The safety turn-off switches on some broilers are intriguing devices. Complex models actually have tiny smoke or flame detectors, which automatically sense burning toast or grease-splattering bacon, and instantly turn off power to the heating element. This is an excellent form of fire prevention. Without internal protection of this nature, broilers can be dangerous appliances if left on while unattended. Toast and greasy meat can burst into live flame that could start a major home blaze.

Cleanliness is not only a preventive maintenance procedure for broiler-type devices, it is an essential safety precaution. The heating elements become red or even white-hot during cooking, and the presence of grease splatters or loose food particles creates a definite fire hazard. After every use, let the appliance cool and then turn it upside down and sideways, shaking out any food particles that may have fallen inside. Depending upon frequency of use of the appliance, every few weeks disassemble the unit for a throrough cleaning.

All electrical appliances carry warnings that they should not be immersed in water. This is because moisture is an electrical conductor and can generate hazardous short circuits, and also some motors and gears may be ruined if they become wet and are allowed to corrode. However, if you use the proper precautions, there are *some* electric heating elements that can be cleaned with water. In these cases the trick is to be sure that they are left open afterward in the direct sun or in a well-aired space and allowed to dry completely before reassembly.

Heating elements in broilers and toasters can be washed in water, *if* the elements can be easily removed from the appliance for separate washing. This can be helpful in instances where there has been a heavy buildup of grease. Be sure to allow the element to dry completely before putting the unit back together and turning it on.

If you are leary of using water, then use one of the commercial oven cleaning solutions. Following the manufacturer's directions, clean the interior of the broiler and the surface of the heating wires or coils on a regular basis. This is not only a wise fire prevention maneuver, but something which will greatly extend the working life of the appliance.

DISASSEMBLY

The on/off switch often has an *interlock* that prevents operation when the door is open. The interlock is a make/break contact switch that is built right into the door. One contact is on the hinged part of the door that swings open; the other is mounted on the appliance frame. With the door closed, contact is made and the on/off power switch works normally. With the door open, the interlock is open, disabling the on/off switch. There are other forms of interlock: the door can have a pin that slips into a slot or hole when the door is closed, and the pressure from the pin activates a "make contact" switch inside the unit. Or the hinge itself may be a part of an electrical switch movement, the contacts opening and closing right along with the door.

In disassembly and in testing for electrical malfunctions, don't forget to check these interlock switches. If they are faulty but you haven't checked them, you could be led by incomplete VOM tests to believe that a more expensive heating element or other component must be purchased and installed. Interlocks are cheap by comparison.

Frequently the heating element in a broiler slides or snaps out the front, just as does the loading tray. The heating element is a series of wires or coils mounted on a metal frame and located just beneath the loading tray. In some models, all disassembly hardware is reached through the opened broiler door. In others, access is by way of a bottom plate or mounting panel at the front of the appliance. You are likely to

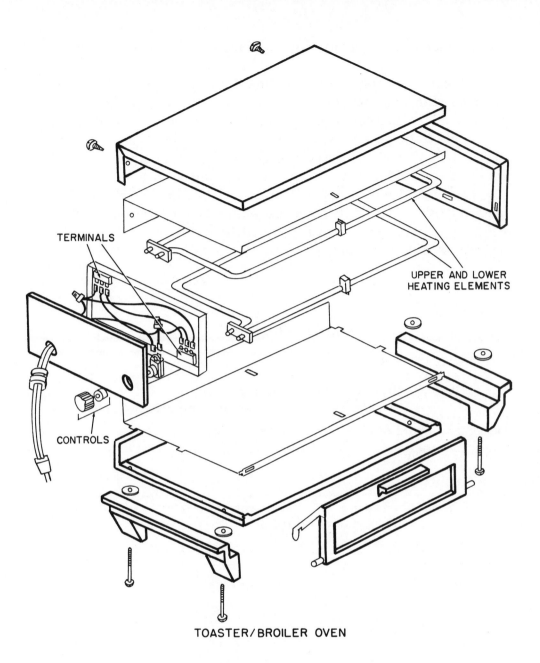

TERMINALS

UPPER AND LOWER
HEATING ELEMENTS

CONTROLS

TOASTER/BROILER OVEN

encounter push-button, rotary and slide switches (lever switches) in this type of device, but you will find that most are of the DPDT type (double pole, double throw — with six contacts). All can be easily tested with the VOM (see Chapters 1, 3, and 4).

The heating element and many of the other components are likely to have plug-type terminals. Occasionally you'll find a unit where the element and components are hard-wired into place (with solder, usually — although soldering a heating element requires special materials since the normal operating temperature of the element will often melt normal solder).

If the broiler cooks but the heating element fails to change temperature according to preset controls, a sensing switch or thermostat is probably at fault. These can be easily checked for continuity with a VOM.

While testing electrical malfunctions in broilers (and in expensive toasters), don't neglect to track down and check all safety switches. They could be the hidden cause of an appliance breakdown.

The limited number of parts makes testing and repair of a broiler oven relatively simple. If power is getting to the oven yet nothing is working, the problem is most likely either in the cord, the switch, the thermostat (if the unit has one), or in the heating element(s).

Troubleshooting Chart

Possible Causes	Solution
Problem: Broiling or toasting fails to begin; machine does not turn on; on lamp does not light.	
Faulty power cord or plug	Check cord and plug
Defective switch	Check switch
No power to outlet	Check outlet
Problem: Breaker or fuse blows when unit is plugged in or turned on.	
Circuit overload	Try another outlet

Can Openers

Most electric can openers are relatively simple in operation. They have a strong metal cutting edge, usually in the shape of a wheel or disc. A motor turns the wheel, and a system of levers and interlocking gears clamp the disc over the edge of the can so the blade can cut through the lid. As the wheel cuts, the motor and gear network rotate the can so that the lid is neatly and completely removed.

For operating convenience, most can openers have lever handles that both clamp the can into position and automatically operate a switch that turns on the motor. Better models come with safety guards to protect fingers from the cutting surfaces and interlock switches to turn off the motor if the machine jams during cutting or when the cutting is complete.

A few older models have an irritating habit of suddenly starting up and running all by themselves while they are supposed to be sitting quietly out on the kitchen counter or in their wall mounts. This is because the operating lever moves up and down, and the trigger spring inside the handle has either weakened or broken completely. With nothing to hold the handle up, the weight of the lever causes it to move down into contact with the motor on/off switch. An expedient cure for such a weakened lever on/off switch is to either unplug the unit when not actually in use or to install a separate on/off switch right in the power cord itself. Or you can disassemble the appliance at the handle to see what part has broken or slipped out of place. Repair or replace the faulty part, and everyone in the household sleeps more peaceably.

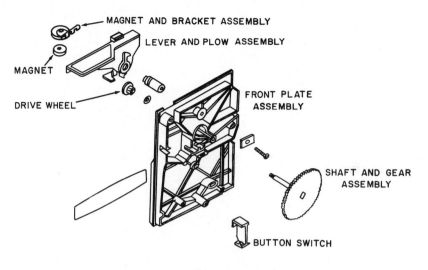

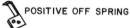

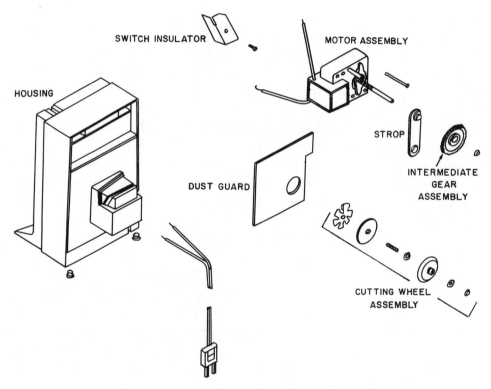

CAN OPENER/KNIFE SHARPENER

Chilton's Guide to Small Appliance Repair/Maintenance
CAN OPENERS

DISASSEMBLY

The cover of electric can openers usually simply slips or snaps off and then back into place. It is held on by either metal spring clips or plastic snap locks rather than screws. Nevertheless, inspect carefully for concealed fasteners, and do not attempt to force open the case beyond applying the steady pulling that may be necessary to open possibly corroded snap openers.

As in all appliance servicing, consult both the operation and maintenance manual obtained when the unit was purchased, if possible. If not, then carefully test and pry and unscrew. Once again, do not force anything!

Care in disassembly is required since many can opener parts involve springs, gears that may be of the slide on and off type, and cambers or tumblers that must be repositioned exactly in order for the unit to work properly. The problem causing your can opener to malfunction could be a simple 50¢ spring or a $1.98 switch. But if careless disassembly causes parts to fly all over the room, you may never be able to put it back together the right way.

Troubleshooting Chart

Possible Causes	Solution
Problem: Can does not properly slide into place; machine does not turn on; on lamp does not light.	
Faulty power cord or plug	Check cord and plug
Defective switch	Check switch
No power to outlet	Check outlet
Safety interlock fails to activate	Check interlock
Problem: Breaker or fuse blows when unit is plugged in or turned on.	
Short circuit	Test
Mechanical jam causing motor overload	Test smooth operation of all moving parts
Problem: Unit turns on, but can does not rotate or cut lid.	
Blown fuse or breaker	Test
Motor winding or lead is open	Test/replace
Connecting wires from power supply to motor or cutting wheel are open	Test
Mechanical binding	Test/replace
Dirt/grime	Clean unit

Chain Saws

The major difference between gas-powered and electric chain saws (beyond the fuel involved) is that gas units have a special clutch that prevents the chain from moving while the motor is idling. When a gas throttle is opened, the clutch engages and drives the chain. The clutch is an advantage because turning a gas engine on and off is awkward and shortens the working life of the engine.

With an electric chain saw, it's no problem to turn the power switch on and off while moving the chain bar from one cut to another, or while changing the angle of a cut. Therefore, no clutch is required in the design of an electric chain saw. (This means the electric models are generally less expensive.)

The disadvantages of an electric chain saw are the need for a voltage source, the awkwardness of a cord (watch it! and don't cut it along with the firewood), and lower power. The gas-powered saws are better for heavy-duty work, and for use in remote locations.

The power unit for a chain saw is not too much different from the motors and switching arrangements for circular and other types of saws. The rotating motor shaft or a chain drive and pulley system drives the chain with its cutting blades around the bar. Maintenance and trouble shooting information are usually well covered in the owner's manual. Just remember that chain lubrication is important for continuous, trouble-free cutting.

The cutting chain is quite similar in construction and design to the drive chain on a bicycle or motorcycle. When replacement is necessary, search for the chain's master link. This will be the one link with a snap-on side plate, while the rest are riveted. Use the blade of a screwdriver or

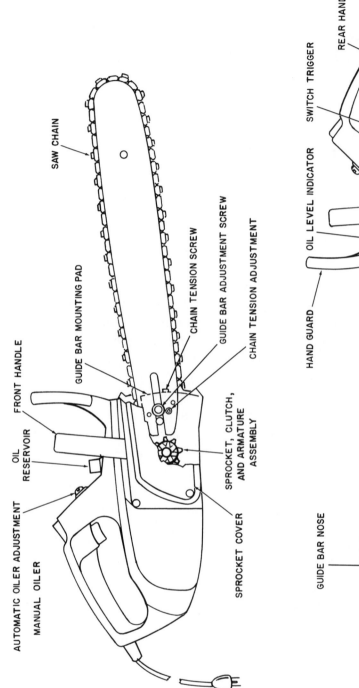

SAW CHAIN

GUIDE BAR MOUNTING PAD

CHAIN TENSION SCREW

GUIDE BAR ADJUSTMENT SCREW

CHAIN TENSION ADJUSTMENT

FRONT HANDLE

OIL RESERVOIR

AUTOMATIC OILER ADJUSTMENT

MANUAL OILER

SPROCKET, CLUTCH, AND ARMATURE ASSEMBLY

SPROCKET COVER

REAR HANDLE

SWITCH TRIGGER

OIL LEVEL INDICATOR

HAND GUARD

BUCKING SPIKE

GUIDE BAR NOSE

GUIDE BAR

CHAIN SAW

pliers to pop off the snap plate. The rest of the link can be easily removed.

Once a master link has been removed, buy a replacement snap plate. Do NOT attempt to use the old one. The slight bending involved in disassembly will make this weaker than the rest of the chain. High torque or snags during tough jobs will cause the weakened master link to snap during operation.

Depending on the size and model of chain saw, the drive may be either direct from the motor shaft, or from a chain sprocket arrangement. Check for excessive sprocket wear, or for sprocket breaks. Either could cause slippage or poor cutting efficiency.

Never attempt to service a unit with the power cord plugged in. Even during bench testing, be sure to put any safety guards or bands back into place. The centrifical force exerted by the moving chain can throw scraps of debris with such force that they can cause severe lacerations to face and eyes.

The instruction book supplied with the saw is the best maintenance guide, as maintenance steps may vary considerably according to make and model. Many modern electric saws have sealed bearings, permanently lubricated. Gear housings require no special attention or treatment in normal use throughout the working life of the saw. Others have plainly marked oil holes, which require the regular application of a dose of light machine oil. Any movable or retractable blade guard should be oiled occasionally to ensure positive action.

As with any saw, the blades should be kept sharp to maintain cutting speed and to prevent excessive friction that could lead to overheating of the motor.

Worn motor brushes in many models can be replaced by removing the brush caps at the motor housing. Because of the construction differences depending upon make and model, write directly to the manufacturer if no instruction book is available.

DISASSEMBLY

The housing of most modern power saws is designed in a "clam-shell" fashion: two halves held together at the seam by bolts or clip fasteners. The power switch is generally mounted as a trigger inside the handle. Be careful in removing the handle from the main housing, for wiring extends from the handle switch down into the housing to the motor. Carefully study the routing of any wiring during break down, so the machine can be put back together properly after repair.

All chain saws have adjustment screws to tighten the chain after installation or after prolonged use (which may stretch the links). Make sure to jot down the placement of any clips or springs on these screws before removing them.

Troubleshooting Chart

Possible Causes	Solution
Problem: Tool does not turn on; on lamp does not light.	
No power to outlet	Check outlet
Faulty power cord or plug	Check cord and plug
Defective switch	Check switch
Problem: Breaker or fuse blows when unit is plugged in or turned on.	
Circuit overload	Try another outlet supplied by a different circuit
Short circuit	Test
Problem: Unit turns on, but does not rotate properly, or tool does not do its work properly; tool overheats.	
Motor winding is open	Test, replace
Connecting wires from power supply to motor are open	Test
Grime/mechanical jam is keeping chain from turning	Clean unit/lubricate
Mechanical binding	Examine unit; repair

Clocks and Timers

Today's clocks and timing devices are often constructed around solid-state quartz crystals, rather than the more old-fashioned gears, levers, mainsprings and electric motors all keyed into movement by a master pendulum or flywheel. Generally, these quartz movements are so inexpensive that it is simpler and cheaper to buy and install a replacement than to repair one. Moreover, many are sealed and virtually impossible to service or repair anyway, other than changing the back-up battery (if it has one).

Consequently, if the clock or timing unit of an appliance goes out, you'll have little choice but to replace the timing unit. The difficulty is that *exact* replacements are usually required. And sometimes that exact replacement is difficult or impossible to find.

In timer mechanisms, the timing device itself is nothing more than a clock. The device is preset by the operator and then automatically activates under certain conditions. A switch of some sort then comes into play. For example, the simple timer of an oven is set by the homeowner, it counts down, and then a switch clicks on a buzzer to let you know that the time is up. A more complex timing device might operate the automatic switching to start the appliance at a preset time, work for a certain amount of time and then shut off later.

The switch in a timer can either be mechanical (such as a gear, lever, or push-pull solenoid) or an electrical on/off switching operation in a solid-state device. Where mechanical parts are involved, servicing is done by disassembling the unit and studying the components. Broken or loose parts, stripped gears, or gears needing lubrication can all be repaired or replaced.

In timers using clockwork movements, it is usually fairly easy to

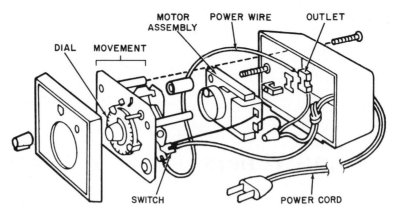

MOTOR ASSEMBLY   POWER WIRE   OUTLET

DIAL   MOVEMENT

SWITCH   POWER CORD

HOUSEHOLD TIMER

Reprinted by permission from *Reader's Digest Fix-It-Yourself Manual* (New York: Reader's Digest General Books, 1977).

spot what is creating the malfunction. Often, cleaning and spot lubrication will correct the fault. The same is true when the timer works but the switching action does not operate. A bent or loose or broken lever or moving part can be the hang up.

When clock-timers rely on electric motors for drive, they are prone to the same disorders as can openers, blenders, mixers and other small appliances that use electric motors. Most clock motors are ac, except for the motors found in appliances that require power supplies to provide dc voltages for the proper operation of other portions of the unit. Examples would be TV sets, radios, video tape machines, record or tape players, and computers. In these, if the timing is not quartz controlled, then the movement is generated by small and inexpensive dc motors.

Whether ac or dc, motor repair is often by replacement. Tests for shorts, open circuits, or continuity are described in Chapter 5. Keep in mind that a dc motor should never be run or tested for any but the briefest periods without being under load. That is, don't hook up the motor to a battery or other voltage supply unless the drive shaft is loaded with the belts, pulleys, or gears it is normally supposed to turn. The speed of a dc motor is determined to a great degree by its load. Without a load, the dc motor can continuously increase speed until it literally self-destructs through heat, vibration, or overwork. The windings can burn open, or the mechanical parts can virtually fly apart, scattering all over the place.

Solid-state electronic clocks are rarely repairable. If power is getting to the unit, yet the unit fails to function, you will almost certainly have to replace the timer as a whole unit.

DISASSEMBLY

Extra precautions are required when removing a timer from an appliance. Timers that are internal parts of TVs, radio sets, or VCRs, may be connected in some way with tuning circuits or gears or with other ancillary functions. You might think a timer does no more than select the time of day a recorder will be turned on or off. However, in some designs that timer also controls the functions of other sections of the appliance. This is especially true in the case of quartz crystal timing devices. Manufacturers may use the capabilities of these tiny units to simultaneously control oscillators (in TV sets, VCRs and microwave devices) and other phase-critical operations.

If your study of the unit indicates that the timer may be integrally connected to other operations, it is better to bring that appliance in to a professional for repairs. If the clock/timer is connected to other critical operations in the appliance, a "retune" of the appliance might be necessary, and this often requires special equipment.

Wherever possible, study the operating or maintenace booklet for the appliance. If you don't have one, make a very careful study of all connections to the timer. As in can openers or any appliance utilizing a "tumbling" movement, cambers or out-of-round shafts and moving parts may be critical to correct function.

Don't accidentally ruin an expensive appliance by trying to fix a timer that is, usually, no more than a convenience add-on. Better to be able to continue to operate the appliance without that extra automatic timing convenience, than to loose it entirely.

Troubleshooting Chart

Possible Causes	Solution
Problem: Timer does not activate as designed; machine does not turn on; on lamp does not light.	
Faulty power cord or plug	Check cord and plug
Defective switch	Check switch
No power to outlet	Check outlet
Safety interlock fails to activate	Check interlock
Timer is jammed/faulty	Repair/replace
Problem: Breaker or fuse blows when unit is plugged in or turned on.	
Circuit overload	Try another outlet supplied by a different circuit
Short circuit	Test
Mechanical jam causing timer overload	Test smooth operation of all moving parts
Problem: Unit turns on, but timing function works improperly.	
Blown fuse or breaker	Test
Motor winding or lead is open	Test/replace
Connecting wires from power supply to timer is open	Test
Mechanical binding	Test/replace
Dirt or grime is keeping machine from working	Clean unit

Coffee Makers

Several types of electric appliances have been designed for making coffee. Some are of the percolator type; others are drip coffee makers.

The simplest coffee makers involve a power cord, an on/off switch, some form of heating element, and a container for the water. More complex models involve thermostats to control water temperature, automatic timers to turn the coffee maker on or off while unattended, supply switches or valves to control the amount of water and/or coffee to vary the strength of the final blend, and automatic cut-off devices to disconnect power from the heating element when the water level becomes low. (A few expensive styles even connect directly to the house water supply. Like coin-operated coffee vending machines, they have a constant supply of drip-grind coffee and water, with switches that refill the blending tanks when fresh brew is desired.)

The most common problem with all models of coffee makers is that they begin to lose their heating power and the coffee is simply not hot enough. This trouble, which is usually caused by a build-up of mineral deposits from the water being used, can be almost completely eliminated through a regular maintenance program.

Whether you use a percolator or drip maker, at least once monthly (and better once weekly) cycle the appliance completely through with a supply of strong vinegar water or a mixture of baking soda and water. There are commercial cleaning solvents available on the market, but either vinegar or the baking soda will work well in dissolving mineral build-ups.

For the vinegar cleaning solution, use a mixture of ½ cup vinegar to a quart of water and put it through the coffee-making cycle (without coffee, of course). When using baking soda, a common mix is two table-

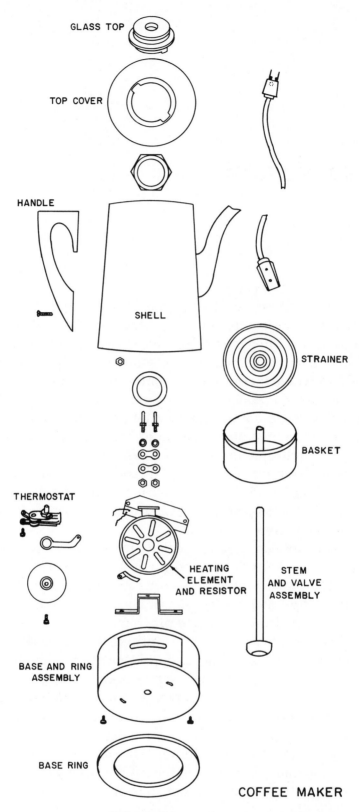

GLASS TOP

TOP COVER

HANDLE

SHELL

STRAINER

BASKET

THERMOSTAT

HEATING
ELEMENT
AND RESISTOR

STEM
AND VALVE
ASSEMBLY

BASE AND RING
ASSEMBLY

BASE RING

COFFEE MAKER

Chilton's Guide to Small Appliance Repair/Maintenance
COFFEE MAKERS

spoons of soda to a quart of water. After the unit cools, rinse it complete-ly. To be sure that you've removed all of the cleaner, it's a good idea to run the unit through a complete cycle with plain water before putting it back into use to make coffee.

A way to greatly reduce these mineral deposits in the first place is to use distilled water in the brewing process. Not only will the de-mineral-ized water reduce the need for such frequent cleaning, but many feel that coffee made from distilled water has a fresher, cleaner taste and they dis-cover that it takes less grounds to make the same amount of coffee.

In cases where the build-up of minerals is so thick that vinegar or commercial solvents can't dissolve them, a steel wool pad or an electric drill with a wire-brush attachment may be used successfully. This is a last-ditch effort, however. The grinding is likely to damage the water container and other components. It can also leave behind harmful parti-cles. After any such cleaning effort, be sure to clean by brewing or perk-ing a vinegar solution and cleaning the appliance thoroughly.

DISASSEMBLY

With most coffee makers, the heating element and many other compo-nents are located beneath the lower cover of the unit. Be sure that the unit has no water in it, unplug it, then turn it on its side. You should be able to locate the holding screws easily. It's possible that one or more screws will have a sticker covering it, with the warning that there are "No User Serviceable Parts Inside."

If the cover doesn't lift away easily, look carefully for screws you might have missed. Many units also have plastic or metal catches to help hold the cover.

The heater element in most units will be plugged or socketed. It's usually a simple matter to replace the element and normally takes just a few seconds. If the element appears to be soldered into place, it is proba-bly special solder that will withstand the heat. *Do not* use regular solder on any heating element.

The "plumbing" is a bit more difficult to get at. Often some of the water-handling components are a physical part of the overall machine. Other parts and tubes are easily replaceable. Fortunately, the most com-mon problem with those pipes (other than getting blocked by mineral deposits) is that one of the connectors comes loose, thus causing a leak. In this case, the simple solution is to slip that connector back into place. In the unlikely event that it has cracked, a replacement will be neces-sary.

Troubleshooting Chart

Possible Causes	Solution
Problem: Brewing or perking cycle fails to begin; machine does not turn on; on lamp does not light.	
Faulty power cord or plug	Check cord and plug
Defective switch	Check switch
No power to outlet	Check outlet
Problem: Breaker or fuse blows when unit is plugged in or turned on.	
Circuit overload	Try another outlet supplied by a different circuit
Short circuit	Test
Problem: Unit turns on, but perk or brew cycle does not begin.	
Blown fuse or breaker	Test
Heating element is open	Test/replace
Connecting wires from power supply to heating element are open	Test
Thermostat is bad	Test/replace
Mineral deposit is keeping heating plate from working	Clean unit
Problem: Coffee maker cycles, but does not keep coffee hot after brewing.	
Thermostat bad	Test/replace
Mineral build-up	Clean
Short in the heating element is causing the unit to operate at a lower temperature	Test/replace
Faulty timing device	Test/replace

Chilton's Guide to Small Appliance Repair/Maintenance
COFFEE MAKERS

Troubleshooting Chart

Possible Causes	Solution

Problem: Appliance leaks, or in a drip-maker the water does not flow from the heating tank into the carafe.

Possible Causes	Solution
Warped, cracked, or broken water tank.	Replace unit
Loose connector	Repair/replace
Clogged pipe	Clean or replace

Problem: Brewing cycle takes too long, coffee is weak.

Possible Causes	Solution
Thermostat bad	Repair/replace
Partially shorted heating element	Replace
Clogged or bent drip or vent hole.	Clean

Deep Fat Fryers

All deep fat fryers use heavy-duty heating elements that can make the oil very hot. This causes a heavier flow of current than with most small appliances, which means that safety precautions concerning cords is more important. Even the right power cord becomes warm on many units. A smaller gauge wire will almost certainly cause problems.

One danger is that operators will use the *wrong* power cord. Don't attempt to use a light power cord from a popcorn popper to plug in your fryer in an emergency. An undersized cord will be forced to carry too high an electrical current. Overheating can lead to shock and fire hazards. When repairing or replacing power cords for these cooking devices, use rubber-insulated power cords, instead of the fiber insulated cords sometimes used for lamps and lesser-duty appliances.

Also because of the current draw of deep fat fryers, overloads of the circuit feeding the outlet are more common. The wattage and current ratings of the particular fryer will tell you if you need to use an isolated circuit to operate the appliance.

Diagnosis and repair of a fryer is the same as with any other appliance with a heating element. Power enters the appliance from the outlet through the cord. If the outlet is dead, or the cord bad, power won't get to the appliance, and therefore nothing is going to happen.

If power gets to the appliance, but the power switch is bad, again nothing is going to happen. If the thermostat or rheostat is frozen closed, the heating element will overheat and cause damage (hopefully just to the appliance itself). If the control is frozen open or fails to make contact, again nothing happens.

The heating element is in series with these components. It is usually housed in the base of the appliance. It can be tested the same way as any

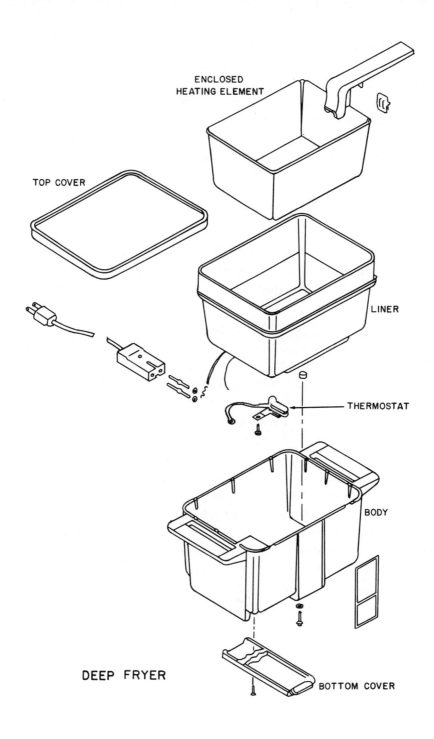

ENCLOSED
HEATING ELEMENT

TOP COVER

LINER

THERMOSTAT

BODY

DEEP FRYER

BOTTOM COVER

other heating element — with a VOM and a test for continuity. The reading should be very low. If the reading is infinity, the heating element has burned or corroded, cutting off the flow of current.

Sometimes the appliance will heat the oil, but not to the proper temperature. This indicates a problem with the thermostat or rheostat, or aging of the heater element. It's also possible that the contacts have become corroded, thus reducing current flow. All can be checked visually or with a VOM, as detailed in Chapter 6.

DISASSEMBLY

The construction of most deep fat fryers is quite similar to that of slow cookers. Beneath the container that holds the fat or oil to be heated is the base. Inside this base are the various components that make the appliance work.

The two parts are sometimes permanently sealed. In this case, attempts at disassembly are futile. If the two *do* come apart, they are generally held together by one or more small bolts that go upwards from the base. A quick visual examination should reveal if disassembly is possible.

Even when disassembly is possible, some units have a completely sealed base. If something is wrong with any of the internal components, the entire base will have to be replaced. This replacement will save you some money, but not nearly as much as you'll save if your particular fryer has replaceable individual parts. Again, once the base is off you should be able to tell very quickly which case applies to you.

Don't force or pry any cover "snap-open" assembly until you are sure there are no hidden screws or clamps. Don't insert screwdriver tips or pointed objects into cases that contain electrical wiring in an attempt to pry a unit open. You may create more internal damage than the minor defect you are attempting to correct.

Troubleshooting Chart

Possible Causes	*Solution*
Problem: Device begins to overheat; unit does not turn on; pilot lamp does not light.	
Faulty power cord or plug	Check cord and plug
Defective switch	Check switch
No power to outlet	Check outlet
Problem: Breaker or fuse blows when unit is plugged in or turned on.	
Circuit overload	Try another outlet supplied by a different circuit
Short circuit	Test
Problem: Unit turns on, but does not cook properly.	
Motor winding or heat element is open	Test/replace
Connecting wires from power supply are open	Test
Thermostat is bad	Test/replace
Grime/dryness is keeping unit from working	Clean unit

Electric Blankets, Heating Pads, and Water Beds

As with any heating element, the wires in the heating element of a heating blanket or pad resists to the flow of electricity, causing the wires to heat up. The most common problems are due to broken (open) or shorted wires, caused as the pad or blanket is twisted and turned by the slumbering user. For this reason, all electrical connections should be checked for potential fraying or excessive bending each time the bed is made. Manufacturers have been successful in developing extremely safe electric heating units, but careless twisting or accidental breaking of power cords is more likely in these than any other type of small home appliance.

Never attempt to repair a shorted, open, or frayed heating element that is sewn inside an electric blanket or pad. A worn wire is likely to have little or no insulation left, and these wires carrying electrical current could then come into contact with a user, leading to severe shock hazards. If a malfunction has been traced to the heating element, discard the unit unless it has a separate element that easily slips inside the plastic or fabric covering.

Repairable breakdowns include thermostat failures, the power cord and plug, the on/off switch, and any temperature control rheostat. The thermostat may be built right into the fabric along with the heating element, but often they are contained in the control unit along with the power switch. Often the power switch is part of the temperature rheostat. In this type of system, the rotor shaft of the rheostat extends through the back plate of the rheostat. There, a camber or lever on the rotor shaft operates a simple SPST switch. The switch part of the rheostat often can be purchased as an inexpensive separate replacement, even if the temperature-control rheostat is relatively expensive.

Test the thermostat with the unit plugged in and the power on. Often turning the rheostat up will cause the bi-metallic thermostat to make a slightly audible click as it activates. Even without the sound to tell if the thermostat is going on, you should be able to determine if the control device is allowing current to pass by testing the temperature of the pad/blanket in a few moments. Similarly, turn the rheostat down, and the deactivated thermostat should make the blanket or pad noticeably cooler in a matter of minutes. If the temperature-control rheostat fails to regulate the heat normally, the problem is usually the thermostat. An obvious indicator of a thermostat problem is when the temperature swings over an uncomfortable range even when the control is set at one position.

The other possible cause of faulty temperature control is the rheostat itself. This may be checked with a VOM. See if the resistance changes smoothly from lowest to highest reading as the knob is turned (with the power off and at least one lead of the rheostat disconnected). If there is a sharp change in the resistance meter reading as the dial is moved, or if the values change in a jerky, erratic manner, the rheostat is bad or dirty. Sometimes it may be rejuvenated by using a commercial electrical spray cleaner, such as is available from Radio Shack or other electronic supply houses. If it can't be fixed this way, then purchase a replacement.

Quite often, the cord/plug and the heating element and blanket are in fine shape, while the controls or thermostat goes bad. It is usually possible to purchase just a new control/thermostat unit from the manufacturer and to splice it into the power cord as a replacement for the defective one.

In water beds, the temperature controls, thermostat, and usually the on/off switch comprise a control unit in the power cord very much like those in electric blankets. Repair and replacement operations are similar. The exception would be those water bed units where the thermostat is placed at the tip of or somewhere within the coils of the heating element. The heating coils of the appliance are placed *underneath* the water bag, on the floor of the water bed cabinet. The bed must be drained in order to obtain access. It is usually possible to test all other parts of the unit first (power plug, cord, on/off switch, and temperature controls) before suspecting either an open heating element or bad thermostat.

The heating element for a water bed may be safely changed — but not really very easily, because you must drain the water bag completely.

Because of the presence of water, special care should be taken dur-

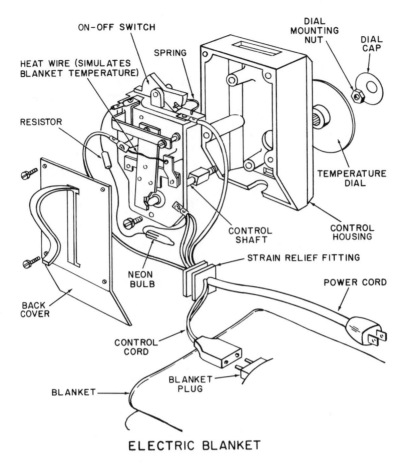

ON-OFF SWITCH

DIAL
MOUNTING
NUT

DIAL
CAP

SPRING

HEAT WIRE (SIMULATES
BLANKET TEMPERATURE)

RESISTOR

TEMPERATURE
DIAL

CONTROL
SHAFT

CONTROL
HOUSING

STRAIN RELIEF FITTING

NEON
BULB

POWER CORD

BACK
COVER

CONTROL
CORD

BLANKET
PLUG

BLANKET

ELECTRIC BLANKET

Reprinted by permission from *Reader's Digest Fix-It-Yourself Manual* (New York: Reader's Digest General Books, 1977).

ing any electrical checks of water beds. Never apply power while any of the electric cables or components or the heating element/thermostat are wet. It's best to totally remove the water bag, and to let the frame dry completely in case of any accidental spills, before testing or repairing the electric heating functions. Then reassemble and fill the bed, with the power disconnected until the bed is ready for use.

DISASSEMBLY

Frequently the screws and other disassembly hardware for a heat control unit are located underneath the knob that is used to vary the temperature or thermostat setting. Most of these knobs simply pull off, though occasionally they are held in place by a tiny set-screw or Allen bolt. Pull off the knob, and the hardware holding together the control unit is generally exposed. Because most of these controls carry fairly high levels of current, they are rarely of the snap-open variety. It's doubtful that an electric blanket or pad would qualify for Underwriter's Lab endorsement without being securely fastened together.

Troubleshooting Chart

Possible Causes	Solution
Problem: Thermostat does not activate as designed; blanket does not turn on; on lamp does not light.	
Faulty power cord or plug	Check cord and plug
Defective switch	Check switch
No power to outlet	Check outlet
Safety interlock fails to activate	Check interlock
Control is jammed/faulty	Repair/replace
Problem: Breaker or fuse blows when unit is plugged in or turned on.	
Circuit overload	Try another outlet supplied by a different circuit
Short circuit	Test
Problem: Unit turns on, but temperature control function works improperly.	
Blown fuse or breaker	Test
Faulty thermostat	Test/replace
Connecting wires from power supply to heater is open	Test

Electric Knives

Gears, levers, and other mechanisms in an electric knife are designed to take the rotary motion of a motor and convert it into the back-and-forth sawing motion of a knife blade. For efficiency and ease of cutting, two or more blades are fastened side by side, with pins and slots allowing the blades to move in opposing directions. As one blade moves outward from the handle, the other blade moves back inward.

The knife contains an electric motor similar to the motors used to drive fans and mixers but designed to operate at slower speeds. There will also be some form of mechanical adaptation to convert the energy from rotational to back-and-forth. The simplest way to develop back-and-forth motion from a turning motor is to mount the motor at a right angle to the blades, with an eliptical camber or gear directly attached to the base ends of the knife blades. The blades, in turn, slide back and forth through machined slots made to hold them in place for proper cutting operation.

The handle contains the on/off switch and often a speed selector. (For speed change in this type of appliance, separate motor windings are generally involved.) The blades usually snap out rather easily, just as beater blades are removed from electric mixers. This is for cleaning and inserting different blades for various types of cutting.

Be sure to follow manufacturer's instructions in the operation of electric knives. Trying to cut in the wrong way or attempting to cut materials that are too tough or thick can swiftly lead to disaster. Next to fans and electric chain saws, these knives are among the most dangerous of small appliances when used carelessly or improperly.

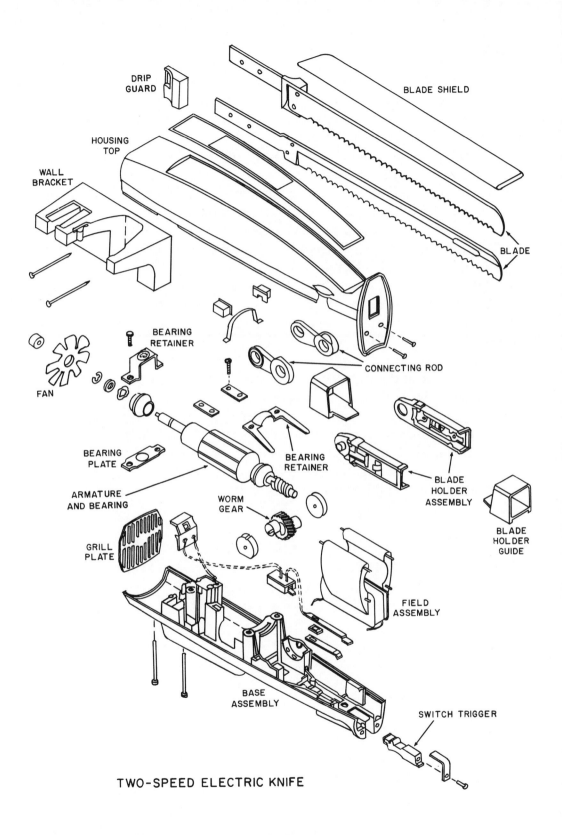

DRIP
GUARD

BLADE SHIELD

HOUSING
TOP

WALL
BRACKET

BLADE

BEARING
RETAINER

CONNECTING ROD

FAN

BEARING
PLATE

BEARING
RETAINER

ARMATURE
AND BEARING

BLADE
HOLDER
ASSEMBLY

WORM
GEAR

BLADE
HOLDER
GUIDE

GRILL
PLATE

FIELD
ASSEMBLY

BASE
ASSEMBLY

SWITCH TRIGGER

TWO-SPEED ELECTRIC KNIFE

DISASSEMBLY

Often the handles are snap-apart rather than screw held, since the knives are meant to be easily disassembled for cleaning and storage. Remove the blades before attempting to take apart the handle for access to the motor and driving cambers/gears.

The components inside may or may not be easily replaceable. Some units have internal components that screw or bolt into place. Others will simply snap in. In both cases, the components and whatever holds them tend to be tricky. You can get into trouble quickly.

Cordless knives use a battery. This battery is usually replaceable, but this sometimes requires desoldering the old unit and soldering in the replacement. Excessive heat can damage the battery. This job is usually best left to a technician willing to make good on mistakes or damage.

The cordless variety will also have some kind of charging unit. If the knife is dead and refuses to take a charge, the fault could be in the knife (probably in the battery), or it could be in the charger.

The charging units are very similar to small power supplies: ac is converted to dc at the right voltage, but at low amperage to provide for a battery recharge. The units themselves are much like the knives. They might be of the screw-together type or the snap-together type. Open the unit very carefully so as not to cause damage, and don't forget to look for hidden snaps and catches. In most cases, repair is by replacement of the unit anyway, so it is rare that you'll need to open the charging unit. (You can test it with a VOM, set to read dc voltage.)

Troubleshooting Chart

Possible Causes	Solution

Problem: Knife handle begins to overheat;
unit does not turn on; on lamp does not light.

Faulty power cord or plug	Check cord and plug
Defective switch	Check switch
No power to outlet	Check outlet

Problem: Breaker or fuse blows when unit is plugged in
or turned on.

Circuit overload	Try another outlet supplied by a different circuit
Short circuit	Test
Mechanical binding may cause motor to overheat	Test freedom of movement; repair/replace

Problem: Unit turns on, but knife does not
vibrate, or blades do not cut properly.

Motor winding is open	Test/replace
Connecting wires from power supply to motor are open	Test
Mechanical binding	Test/replace
Blades are dull or are wrong for desired cutting operation	Sharpen/replace
Grime/dryness is keeping motor from working	Clean unit/lubricate

Fans

An electric fan is nothing more than an ac motor driving a rotor shaft, to which is attached some type of blade to move air. Refinements in complex models might include gears, cambers (eliptical wheels), or levers designed to move the direction of the fan up and down or from side to side. Some fans have speed switches to change the speed at which the fan blades rotate, and some even have thermostats to determine when fans should be turned on or off.

Fans are often built-in components of many other appliances. In electric heaters, fans are used to blow heated air into a cool area. In other appliances, they are used to dissipate heat out of an appliance so the heat won't damage other sensitive electrical components.

Room fans to circulate air are obnoxious when they are noisy. The racket from a rattling fan is caused either by the fan blades hitting some part of the protective cage or by the vibration of loose parts. Overheating or lack of lubrication can also cause the moving parts of the motor to develop friction and noise. To determine the cause of irritating sounds from a fan, listen closely. Turn the fan to various speeds, if it has such a control. Tilt it slightly from its normal operating position. Turn it off and rotate the blades, to see where they might be hitting some obstruction. Look for nicks or scratches, which may indicate where a vibrating blade is striking a cover or wire safety cage. A blade may move freely when spun by hand, but some vibration or wobbling could cause it to hit a random surface when the motor is again turned on.

Cleaning and lubricating is the key to long life and efficient fan operation. Full instructions on the maintenance and test/repair of electrical motors are given in Chapter 5 of this book.

It is wise to clean and lubricate *all* electric motors and driving

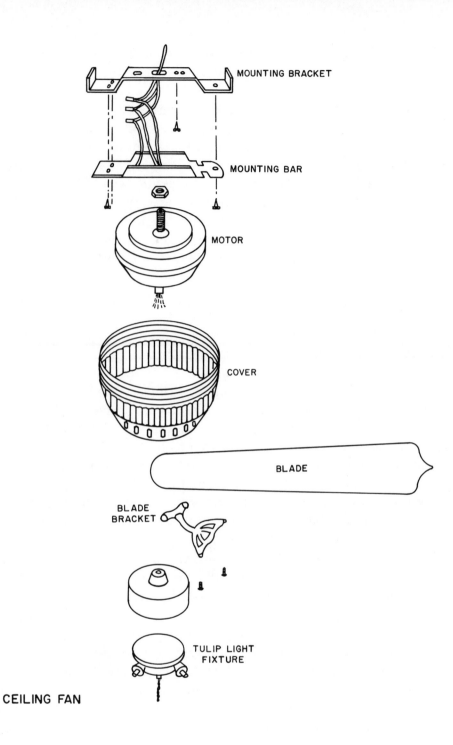

MOUNTING BRACKET

MOUNTING BAR

MOTOR

COVER

BLADE

BLADE
BRACKET

TULIP LIGHT
FIXTURE

CEILING FAN

mechanisms in small appliances at least once every six months. The maintenance schedule should be more frequent for appliances used most often.

Methods of changing the speed of fans varies. Some fan motors have separate windings; voltage is supplied to one winding for one speed and then switched to a second winding for another speed. Some fans have a lever that moves either the motor or a variable-diameter drive shaft, so that speeds are determined by the diameter of the part of the shaft in contact with the motor. In fans driven by dc motors, the speed switch could control the voltage supply to the fan. Each system involves a separate system of testing, repairing, or replacement. Be sure you understand the functions of the speed selector system before attempting repair.

DISASSEMBLY

Just how the fan comes apart will depend on how it is built. Large box fans have protective covers front and back that can usually be removed for access. Table models often consist of a base containing the switch and other controls, plus an upper housing to encase the fan motor. However it's put together, and whatever the order, the basic construction consists of the fan motor, the blades, and the switches and controls. The individual cases are usually held together by bolts, although a few of the less-expensive models have plastic cases that are glued together.

Once inside, it is usually a simple matter to trace the various wires. Each of the components can be tested with a VOM, although this will sometimes mean that you will have to unsolder a wire before the continuity test can be performed accurately.

Take care in disassembling the unit, and be sure to have your notebook and sketch pad at hand. Be sure that you determine whether the fan motor mount is rigid or floating as you take the fan apart, so you can reassemble the device properly.

Do not apply power with the safety cage removed. Spinning fan blades have taken the fingers of hundreds of careless users. Another good precaution is to wear safety glasses or goggles when testing or repairing fans, since the rapidly spinning blades can easily blow particles of dust, metal, or other irritants into the eye. For this reason, it is also advised that you not work on fans while pets or small children are around.

Troubleshooting Chart

Possible Causes	Solution

Problem: **Fan cooled device begins to overheat;**
fan does not turn on; on lamp does not light.

Possible Causes	Solution
Faulty power cord or plug	Check cord and plug
Defective switch	Check switch
No power to outlet	Check outlet

Problem: **Breaker or fuse blows when unit is**
plugged in or turned on.

Possible Causes	Solution
Circuit overload	Try another outlet supplied by a different circuit
Short circuit	Test

Problem: **Unit turns on, but fan does not rotate properly,**
or cool air does not flow properly.

Possible Causes	Solution
Blown fuse or breaker	Test
Motor winding is open	Test/replace
Connecting wires from power supply to motor are open	Test
Thermostat is bad	Test/replace
Grime/dryness is keeping motor plate from working	Clean unit/lubricate

Garage Door Openers

Electrically operated garage door openers are generally for overhead doors that slide up and down on metal tracks. The mechanical setup is a bit different in one-piece doors that glide in straight tracks than in sectional doors that move in curved tracks. The greatest problems with these units often come during installation, where the physical size of the door opening (height and width) determine the size of the control unit required to drive a particular door, or from a poor installation. The installer may discover that an undersized control unit has been purchased, or will learn that the "square" opening is far from true.

If the door opening is two or three inches different in top width from bottom width, the frustrated installer is going to have to learn a great deal about the proper way to mechanically adjust the tracks and all levers, brackets, drive arms, clutch wheels, and opening/closing adjustment nuts. The less perfect the installation, the more problems there will be down the road.

For those who encounter operational failures in doors they did not personally install, it is best to make a thorough study of the installation/operation instructions that came with the door opener, and to make a complete study of the installation. In instances where your own unit is binding (not opening or closing completely) it might be advisable to inspect a friend's properly working unit in order to observe the exact moves of the mechanical parts and clutch operations so as to better understand what may have gone wrong with your own.

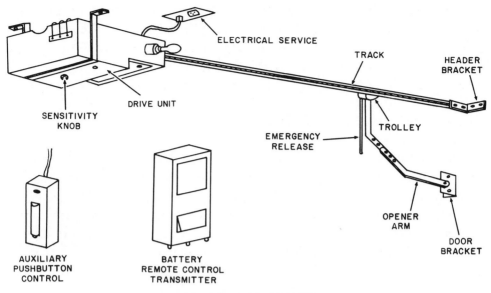

ELECTRICAL SERVICE

TRACK

HEADER
BRACKET

DRIVE UNIT

SENSITIVITY
KNOB

TROLLEY

EMERGENCY
RELEASE

OPENER
ARM

DOOR
BRACKET

AUXILIARY
PUSHBUTTON
CONTROL

BATTERY
REMOTE CONTROL
TRANSMITTER

GARAGE DOOR OPENER

MAINTENANCE AND ADJUSTMENTS

Failures come under four general categories:

1. The remote-control or the direct manual switching device fails to make the door operate.

2. The door jams part way open or closed or causes a great deal of vibration or noise as it is operated.

3. An overload, motor breakdown, or electrical failure has caused the unit to become totally inoperable.

4. The unit operates too slowly or weather conditions cause unreliable or intermittant operation.

The most common difficulties arise from improper maintenance. Even though most modern units track on silicon-lubricated sealed bearings and the control motors are self-lubricating sealed units, the need for regular oiling, clutch wheel adjustment, cleaning, and mechanical tightening of all moving parts and brackets is essential for consistent quality perfomance. Garages are more prone to dust and grime build-up than the interiors of most homes. Use a blower or a slightly damp soft cloth to gently wipe dust and excess grease from all moving parts. In particular, keep the motor assembly housing clean, for a covering of dirt can lead to overheating and damage.

If your unit contains an exposed pulley and belt drive, make sure this drive system is clean and the tension adjustment is correct according to the owner's manual. If there is slipping or wear in the drive belts, replace them. A little extra life can be given to drive belts by properly readjusting them and by dressing the surface that runs over the pulley and drive wheel with soap, parrafin, or bee's wax. With the motor off, wipe the inner surface of the belt with the dressing, manually rotating the drive so that the entire inner belt is coated. Use a soft cloth to wipe off any excess. Then start the motor. Let the drive run in normal operation for 2 or 3 minutes, then adjust the tension according to the instruction manual.

Your door opener may require occasional adjustments to make it close more snugly or to get it to open wider. The operator's manual will describe a clutch adjustment to permit the door opening/closing drive to slip if the movement of the door is somehow obstructed. This slippage is a safety factor, preventing any possible injury to a person or pet caught accidentally in the closing door. When the obstruction remains for more than a few seconds, the automatic reverse control of most automatic door-opening devices takes over and re-opens the door.

Opening adjustments must be made with the door *closed*. Turn most adjustment controls to the right if the door is not opening wide enough and to the left if it is opening too wide.

Make the closing adjustments with the door *open*. Turn the control to the right to close less tightly and to the left to close tighter.

Generally the automatic reverse mechanisim can be adjusted to make the door reverse its direction of movement under minimum pressure. The reverse operation is usually controlled by a spring-loaded switch. For maximum sensitivity, the adjustment screws should be turned so the gap in the spring is about ⅛ of an inch.

Clutch wheels should be adjusted so as to move the door through its proper cycle, but be loose enough to slip if there is any obstruction to the movement. Test the slippage and reversal mechanisms by putting an empty cardboard box between the doorframe and the closing door. If the automatic device slips and/or reverses without damaging the carton, the set-up is proper. If the box is crushed before the reversal takes over, the clutch belt assembly is probably too tight.

Whenever working on any garage door opener, keep safety first and foremost in mind. A garage door is heavy, and the various springs and other mechanical parts are built to handle a lot of weight. People have been severely injured by parts that are suddenly released from tension.

Before beginning, read through all of the literature, both on the door and on the opener. Study all parts and try to determine where there are dangers.

A WARNING ABOUT FCC REGULATIONS

Some people know how to increase the power of radio-controlled devices using illegal boosters. For example, some people have used illegal boosters to increase the power of Citizen Band radios in order to permit communication over longer distances. A few homeowners, wishing to operate their garage doors from longer distances, have used similar power boosters. This is wrong and dangerous.

The power of radio-controlled devices for automatic garage door openers is restricted and licensed by the Federal Communications Commission. One San Francisco homeowner is now serving a lengthy prison sentence for manslaughter, after having caused the deaths of seven persons. Without knowing it, when he boosted the power of the remote control device for his garage door opener he generated a subharmonic (sub-multiple) of the unit's operating frequency. This high-powered illegal subharmonic happened to fall into the same frequency range as the control tower radio frequency of a nearby small airport. A private plane was attempting an instrument landing in heavy fog. At a point less than 100 feet from the ground, interference from this owner's modified garage door opener blanked out all radio and direction-finding equipment on the aircraft. The plane crashed into an airport building. Seven were killed, but the FAA was able to recover a flight recorder from the damaged aircraft. It took over four months for the FAA officials to trace the repeat interference on this same frequency, but the man who modified his garage remote control device was eventually located, arrested, tried, and convicted.

The safety warnings on all electrical devices are placed there for solid reasons, *even if the home repairman does not fully understand their need*. Never tamper with devices in an illegal or unsafe manner.

Troubleshooting Chart

Possible Causes	Solution

Problem: Door fails to open or close when remote control device is used, but does operate correctly when operated by indoor manual control switches.

Possible Causes	Solution
Faulty remote device	Check battery and remote control switch.
Bad transistor or other component in remote control device	Service the unit or take to a professional; replace

Problem: Remote control device operates door, but direct manual switch fails to cause door to close or open.

Possible Causes	Solution
Switch is not set close enough	Adjust
Defective switch, or wiring open(short).	Check/replace switch; check wiring with VOM

Problem: Door opens or closes only part way, no matter whether operated by manual or remote control switches.

Possible Causes	Solution
Obstruction	Check for proper adjustments, lubrication, or mechanical jamming. Could be caused by bent tracks, loose mounting brackets, slipping clutch, or any physical binding of door movement. Disconnect connecting arm of automatic door opening unit and manually open and close the door to see if it is tracking properly without being driven by the electrical drive motor.
Slipping	Clutch adjustment, or replacement of drive belt, if any. Check to see that the track has not become loose or misaligned. Dress and readjust belts.

Troubleshooting Chart

Possible Causes	Solution
Problem: Motor turns on, but open/close mechanism fails to operate properly.	
Lubricant	Lubricate unit and moving parts
Ice buildup in track, or moisture freezing the door shut	Warm garage or track; clean. Disconnect until warmer weather arrives.
Problem: Electric motor fails to operate at all.	
Electric failure	Check breakers, source voltage supply, and all switches (including reversing switch)
Motor failure	Check and repair/replace
Problem: Fuse blows or breaker trips when device is operated.	
Short or motor overload	Check all switches, motor, and wiring with VOM
Circuit overload	Check circuit, or use another
Door binding; slippage adjustment too tight	Adjust

Grass Trimmers and Lawn Mowers

There are a number of electric lawn care appliances, the most common of which is the grass trimmer. All electric lawn care tools consist of basically four parts: the casing, the on/off switch, the motor, and some kind of cutting blade or string. All operate on similar principles.

If the lawn appliance doesn't work at all, you have three places to look. There may be no power (or insufficient power) getting to the appliance (a dead outlet, or an overly long and/or overly small-gauge extension cord). This is easily checked by using a VOM first at the outlet and then at the far end of the extension cord.

Within the appliance, a bad switch could be keeping power from getting to the motor. Or the motor could be bad, in which case getting power to it won't help. Again you can test with a VOM. First test for continuity across the switch. (Many units allow this with a minimum of disassembly.)

You can also test the motor itself (following the tips in Chapter 5) to see if this is the problem.

Sometimes as you're disassembling the tool for these tests, you'll discover some other source for the problem, such as a drive belt that has slipped off or broken.

More commonly, the appliance works but makes a funny sound. This is usually caused by blades that have clogged or jammed. The electric motors used in these appliances are usually small and not very powerful. Shut off the switch, then unplug it. Now you can safely turn the tool over to investigate. If the motor and/or blade is clogged, clear away the debris before you go back to work. That overload on the motor can burn it out very quickly.

Grass trimmers use a nylon or plastic cord to cut. Especially if the trimmer has an automatic feed, the cord may have worked its way out too far and become tangled. Again, a quick visual check will tell you if this is the case.

Electric lawn mowers are less prone to clogging than grass trimmers. The blades are either metal or of relatively non-flexible plastic. The torque of the motor also tends to be higher, making it easier for the blades to slice through various plants. Even so, any lawn mower can find itself in difficulty if the plants are too heavy or are too wet, and especially so an electric lawn mower. Vibration and extensive use also leads to wear and breakage of many mechanical parts.

In all mower servicing and maintenance procedures, make sure the electrical power is turned off and the power cord is disconnected. Those blades are highly dangerous if turned on accidentally while the machine is on its side or upside down.

Check all nuts, bolts, and screws before each use of the mower. In particular, make sure that the bolts holding the cutting blades are tightened to about 50 ft-lbs of pressure.

Some operators simply remove any damaged guard, safety device, or protection flange. After all, the unwise operator reasons, the part had no effect upon the smooth running of the machine. But danger from rocks, sticks, dirt and other flying debris kicked into the air by the swiftly moving cutting blades is immense. *Do not attempt to run your mower without all safety guards and devices in place and in good working order.*

The appliance should be cleaned thoroughly after each use. If possible (and safe), tilt it on its side and use a water hose to spray the underside of the mower housing and the cutting blades in order to free any stuck masses of grass clipping. (Be sure to read the owner's manual first to see if you can safely use water to wash the tool.) Then use a soft dry cloth to wipe clean all of the upper surfaces.

Allow the motor to cool before storing your mower in any enclosure. The housing could get hot enough to set fire to any nearby oil-soaked rags or cardboard containers.

After every 30 or 40 hours of operation, and at the end of the lawn mowing season, apply two or three drops of light machine oil on the inside of all wheel bolts. Spin the wheel to distribute oil into the wheel bushings. At the same time, check to see if there are regular lubrication points or holes for the electric motor.

Also check the cutting blades to see that they are tight and to determine whether they need resharpening. If you use a file and handle your own resharpening, it is wise to purchase a blade balancer (available at

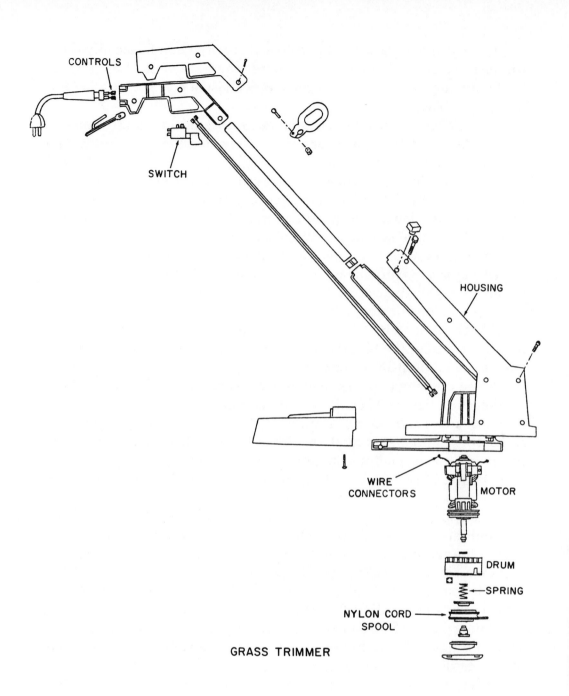

CONTROLS

SWITCH

HOUSING

WIRE
CONNECTORS

MOTOR

DRUM

SPRING

NYLON CORD
SPOOL

GRASS TRIMMER

Chilton's Guide to Small Appliance Repair/Maintenance
GRASS TRIMMERS AND LAWN MOWERS

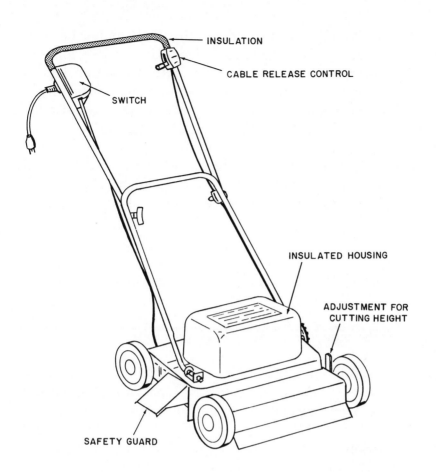

INSULATION

CABLE RELEASE CONTROL

SWITCH

INSULATED HOUSING

ADJUSTMENT FOR
CUTTING HEIGHT

SAFETY GUARD

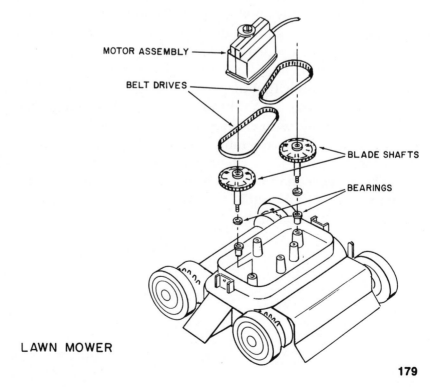

MOTOR ASSEMBLY

BELT DRIVES

BLADE SHAFTS

BEARINGS

LAWN MOWER

most hardware stores) so that you can determine if the cutter has suffered abnormal wear at one end. A balanced blade will stay in a horizontal position when placed on the balancer. By contrast, the unbalanced blade will tilt to the heavy side. In the latter case, file more metal off the cutting edge on the heavy end of the blade, then check the balance again.

Whether or not the lawn tool has blades, check the other mechanisms for wear and tightness.

Be sure that the lawn care tool you are using is meant for the job at hand. Misuse is the single greatest enemy of lawn care appliances, following closely by poor storage and poor cleaning habits.

DISASSEMBLY

Steps for disassembly depend on the appliance. Most consist of three basic sections.

The section closest to the ground, where the work is being done, normally contains the motor and cutting blade(s), along with most or all of the mechanical parts.

The topmost section contains the various switches and controls, and is generally where power enters the tool. In most cases, there is not much more than the incoming wires, an on/off switch, and the outgoing wires to the motor.

The middle section, the handle is usually hollow to allow passage of the wires. Only rarely will you find anything else in this middle section.

A special case is an electric leaf blower. It consists of a fan with a nozzle (usually flexible) coming from it. The motor and fan blades are most often in the top section. This section also contains the switch and any other controls that are used on that model.

To disassemble, first inspect the casing. The screws for disassembly are usually obvious. In addition, most outdoor appliances have snaps and catches, along with various pressure fittings. These help to keep dirt out, but can make disassembly more difficult. If you take your time and don't force anything, you'll soon figure out how it all comes apart.

The lower section will have the various mechanical parts, including various springs and clutches. Careless disassembly can make reassembly difficult or impossible.

Troubleshooting Chart

Possible Causes	Solution
Problem: Appliance fails to work. Nothing happens.	
No power	Test outlet; check fuse or breaker
Bad switch	Test switch; replace
Bad motor	Test; replace
Short in wiring	Test for continuity; repair
Problem: Strange noises; motor stops.	
Mechanism is jammed	Inspect and clean
Blade bolt is loose	Tighten to 50 ft-lb
Problem: Unit turns on, but does not run smoothly, or does not cut grass properly.	
Overheated bearings	Test/lubricate
Motor winding or switch element is open	Test/replace
Connecting wires from power supply are open	Test
Drive belts are bad	Test/replace
Grime/dryness is keeping unit from working	Clean unit/lubricate
Problem: Mower vibrates abnormally.	
Blade bolt is loose	Tighten
Cutter blades unbalanced	Sharpen and check balance

Hot Plates

Hot plates belong to the same breed as the warming element in your automatic coffee maker, an electric skillet, slow cooker, and electric deep fat fryer.

In the heating element of a hot plate, electricity flows through the wire of the element, causing heat to be generated. A rheostat governs the amount of current, and thus the amount of heat. This rheostat control is usually calibrated with various temperature settings. In most hot plates, the rheostat also serves as the on/off switch.

In some cases, there will also be a thermostat. This senses the temperature of the element and cuts off the current flow if it gets too hot and then cuts it in again as the temperature drops. The thermostat keeps the plate of the appliance at a constant temperature.

Both the rheostat and the thermostat are usually contained within the same control box. Sometimes they are separate, with the thermostat more of an integral part of the heating element.

Occasionally touching a hot plate gives the operator a shock. This means that somewhere inside a voltage-carrying line is being shorted to the chassis. Under no circumstances should the appliance be used until the cause is tracked down and fixed. First use your eyes. Look for melted wires or spilled food that could have caused a short to the case.

You can then use your VOM to test for continuity (or lack of continuity) of all wires. There should be a reading of infinity between the "hot" wires and the metal chassis. If you get a reading, continue your search until you find out what is touching the case that shouldn't be.

Some hot plates are prone to damage from spills. Wipe up all spills when using the hot plate, and clean it regularly. (If you've ever had to scrape through a half-inch layer of grease and dried, hardened food to get at the contacts of a switch, you'll know how much easier regular maintenance is than repairing the resultant damage.)

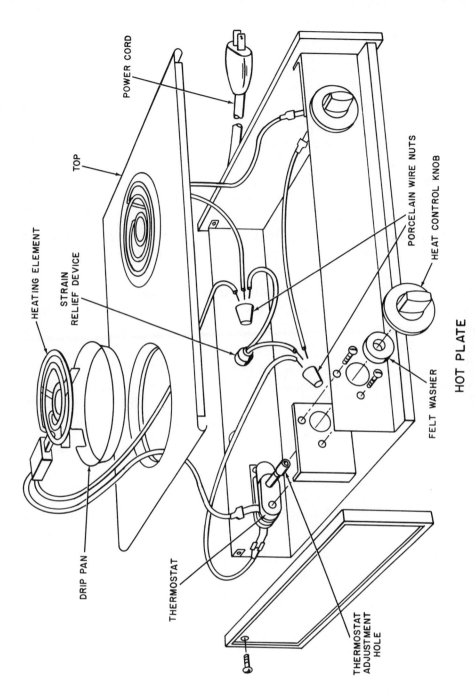

POWER CORD

TOP

HEATING ELEMENT

STRAIN
RELIEF DEVICE

DRIP PAN

THERMOSTAT

THERMOSTAT
ADJUSTMENT
HOLE

PORCELAIN WIRE NUTS

HEAT CONTROL KNOB

FELT WASHER

HOT PLATE

Reprinted by permission from *Reader's Digest Fix-It-Yourself Manual* (New
York: Reader's Digest General Books, 1977).

183

DISASSEMBLY

The way to disassemble the unit depends on how it is built. Some have exposed spiral elements on top, much like an electric range. Normally these can be easily tested and replaced, as they usually just plug into a socket. A few will have lugs, and they are also easy to replace.

Some models have a smooth plate, much like the warming plate of a coffee maker. In this case, the heating element is generally accessed from beneath. Unplug the unit and turn it over. You will be able to spot the holding screws easily enough. Be sure to examine the appliance first to make certain that you don't unscrew some internal component.

In other models, the unit is held together from the top, or from the sides, or both. In almost all cases, the way to open the case is usually obvious.

Once inside, the heater element is generally easy to replace — and also easy to disconnect for testing. The other components are also tucked inside the case and are accessed from beneath. Usually there is plenty of room inside for probing and testing.

The first thing to do whenever opening this or any other appliance is to visually examine everything. Look for the obvious, such as corrosion, burns, and other damage. It could be that the entire problem with the appliance is nothing more than a corroded contact.

Troubleshooting Chart

Possible Causes	Solution
Problem: Won't heat; totally non-operational.	
No power to outlet	Check outlet
Power cord is bad	Test for continuity; repair/replace
Switch or other control is bad	Test; repair/replace
Heating element is bad	Test; replace
Problem: Improper or no control of temperature.	
Thermostat, rheostat, or control device is bad	Test; replace
Heating element is aging	Replace

Irons

Most common problems in irons occur from defective cords, plugs, or switches, improperly operating thermostats or heating elements, and plugged openings or safety valves in steam irons. Automatic irons also have mechanical parts that can break or malfunction.

When replacing the cord, keep in mind that the cord must draw a large amount of current. For a replacement cord, be sure to pick one that is capable of handling the load.

The thermostat of your iron will generally be in one of two places. On some irons an adjustment screw is located externally in the handle and can be adjusted without disassembling the iron. On others the only visible control will be the normal temperature setting control, with the actual adjustment being made at the body of the thermostat. This means that to fix the thermostat you would at least have to remove the handle of the iron.

A visual examination of the thermostat might reveal obvious damage, such as burning. If this is the case, it's time to replace that thermostat if possible, or the entire iron. Occassionally you can bring your thermostat back to life by cleaning the contacts of the thermostat, such as with a file used to clean the points on an automobile.

The heating element is most often sealed inside the base of the iron. Only rarely can you replace just the element; and in some cases you can't even replace the base. As a general rule all you'll be able to do is to test the heater element for continuity (see Chapter 6) to find out if the element is causing the trouble, and then replace the base, or the iron if this isn't possible.

Steam irons are more complex, and there are more things to go wrong. Preventative maintenance is the best way to avoid problems. Use

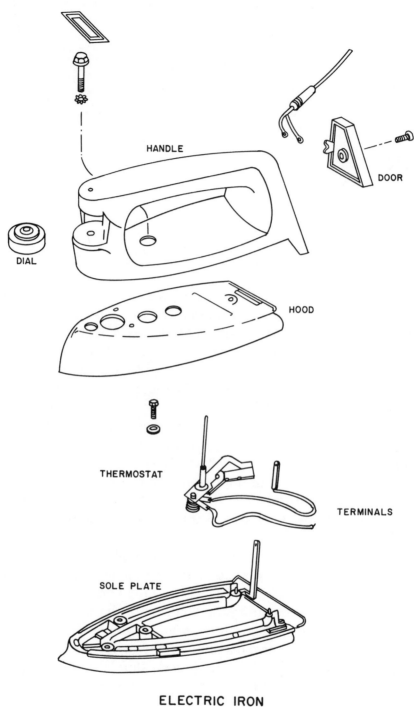

HANDLE

DOOR

DIAL

HOOD

THERMOSTAT

TERMINALS

SOLE PLATE

ELECTRIC IRON

only distilled water. This will greatly reduce build-up in the tubes, valves, and outlets. If the outlets are clogged, you can sometimes clean them with a pipe cleaner. Do so carefully, as it is easy to cause damage.

DISASSEMBLY

The casing of an iron is usually a heavy, solid unit, which opens only from the top. Disassembly screws and clamps are generally found in the area of the handle, and are often hidden (for cosmetic reasons) beneath the trim of the iron. On irons with a filler spout for water, you often have to remove the cap before the iron will come apart.

Proceed carefully when taking a unit apart, because often small springs and tiny bushings are involved. These may fall or snap out before you have observed the exact way they are to be replaced when the device is reassembled. Make pencil drawings as you take an iron apart, marking down the order in which washers, bushings, electrical contacts, etc., are placed on the shaft of a bolt. Follow wiring color codes and guides as you remove switches, temperature control rheostats, or thermostats, in order to properly replace them.

The more features your iron has, the more complicated it will be to disassemble (and reassemble). Irons with water reservoirs for steam present some special problems. Along with the heat gaskets and heat shields, all water-carrying tubes and reservoirs must be sealed, and the sealant must be able to withstand heat.

Chances are very good that you'll damage or destroy gaskets and seals in disassembly. Using the correct sealant in reassembly is important. Some sealants will not last under the conditions inside an electric iron and will actually cause new hazards. A poor seal that allows water to drip into and onto the electrical contacts, for example, can create a serious danger of shock.

Problems with plugs, cords, and switches may be traced and corrected using the electrical wiring tests and troubleshooting advice of Chapters 2 and 3. Heating elements and thermostat control devices are discussed thoroughly in Chapter 6. Both should be read and studied before attempting repair of any home iron.

Troubleshooting Chart

Possible Causes	Solution
Problem: Iron fails to turn on or heat up.	
No power	Check outlet
Overloaded outlet	Try another outlet
Faulty switch, plug or cord	Test; repair/replace as needed.
Problem: Pilot light goes on, electricity is getting to iron according to VOM test, but iron does not heat properly, or temperature is too low.	
Bad heating element	Test/replace
Faulty thermostat	Repair/replace
Faulty heat control element	Repair/replace
Problem: Iron heats, but no steam comes out for pressing.	
Clogged vents, valves, or feed tubing	Clean mineral deposits
Improper water feed	Check water feed; repair/replace

Juicers

In mechanical movement and design of operation, a juicer is unique. But in the matter of component parts (electrical), it is quite similar to many styles of food processors (see pages 127 to 132). The juicer has a spinning or whirling scraper that is inserted into half of an orange, lemon, lime, etc. When the unit is turned on, the scraper scrapes all the moisture and pulp from the fruit into a holding tank or pouring container. Some models have strainers to separate seeds and solids from the juice.

Juicers have electric cords and plugs, switches (on/off and speed and/or size controls), and an electric motor. Generally the motor drive shaft is attached directly to the fruit scraper, but a few complex models have belts and/or speed-change gears. Because of the force involved when a juicer is turning at high speed, accidental jams may lead to bent parts—even to a bent motor shaft. Check for mechanical binding when the motor shaft or the scraper does not turn easily when the juicer is disassembled (with the power off and plug disconnected). If something is making the rotating motion stop, a careful inspection will probably reveal the bent or broken part. With some less-expensive models, the entire scraper is made of plastic. The shaft beneath might have worn away enough of the scraper so that the shaft spins merrily along, without propelling the scraper.

Food residue can create motor and speed problems. Frequent cleaning of the appliance is recommended. About once every month (more often if used daily), disassemble the unit and clean all parts thoroughly according to instructions in the owner's manual.

DISASSEMBLY

First carefully remove and set aside any glass or ceramic parts, such as the juicer or mixer bowls or pitchers.

Countertop appliances have most of their screws and bolts on the bottom plate, although you will occasionally find screws tapped downward near the motor shaft, under the pitcher/bowl location. Be careful before removing any of these, for they may be motor- or switch-mounting screws instead of disassembly screws. If disconnected with the bottom plate still on, the motor or switch could drop down out of place, resulting in damage or lost parts.

In countertop juicers the motor is usually in the base, and gears or a drive shaft extension reaches up to drive the scrapers.

Some baseplates have snap-on fasteners, which may be popped or lightly pried loose by hand or with a screwdriver. Don't force anything apart until you have made a thorough search for hidden screws or clamping fasteners.

Study the motor mount arrangement. Speed changes can be effected by changing the physical position of the motor, with the drive shaft ground into different circumferences at the points where it is moved to drive the appliance. Other types of speed controls include sets of gears, or switching changes that actually vary the turning speed of the motor.

Move bearings, bushings, gears, and drive shafts gently to learn the mechanical actions of the appliance for proper operation. This will be of help in correcting any jamming or binding difficulty.

Troubleshooting Chart

Possible Causes	Solution

Problem: Machine fails to start; blows fuses/breakers; pilot light fails to light; on/off switch does not work.

Wall outlet, power cord, plug, power switch, fuses/breaker, internal reset button, etc.	Check/repair/replace

Problem: Voltage at switch ok, but motor does not energize.

Bad switch or connector	Check; repair/replace
Faulty motor	Test; repair/replace

Problem: Appliance works, but difficult or impossible to change speeds.

Speed switches	Test; repair/replace

Problem: Can hear motor turn on but nothing moves.

Binding of some part	Disassemble; test
Motor jammed	Test, lubricate; repair/replace

Problem: Appliance seems to be working, but juicing operation/function is not satisfactory.

Improper scraper head	Check manual to see that right blade/scraper is used.
Broken or bent parts	Repair/replace

Power Tools

Successful repair of power tools often depends on having the right repair tools on hand. For example, many rotating shafts and gears on electric saws, drills, sanders, etc., are held in place with small allen-type set screws. These allen screws may be metric or American sizes. There are two types of allen wrenches — one for metric and one for the American standard. An American standard allen wrench may *seem* to fit the allen screw on the drive shaft of an electric drill. However, close inspection may reveal that the fit is not perfect, and a tiny bit of metal is worn away from either the end of the wrench or the fitting on the top of the screw each time it is turned. Eventually, either the screw or the wrench will be "stripped," and become too loose in its fitting to be turned at all. The allen set screw will have to be drilled out with a "handy-out" attachment on yet another electric drill, with the hole then retapped for the next larger size allen set screw. The same can occur when working with standard screwdriver slots, or with Phillips screws. Use the *right* tool head for the screw, and you will prolong the life of the device.

Use the wrong tool, or a convenient "substitute," and you are heading for a world of trouble. A repair job that should have involved the simple replacement of a 50¢ allen screw winds up being extremely costly due to inefficient and incompetent home repairs.

Most home workshop power tools are driven either by a rotating electric motor (drill, saw, sander, etc.), or by some form of impact twisting or pounding motion (electric automobile lug wrench, electric screwdriver, electric hammer, etc.). A careful review of Chapter 5 on electric motors and their maintenance or repair is in order before attempting to fix any broken power tool.

Doubly important are commonsense electrical and mechanical safety precautions. Avoid working while standing on damp floors or damp earth. Wear rubber-soled shoes. Watch fingers and hands to avoid nasty cuts. Never work on power tools while they are plugged in. All too often this basic rule of safety is forgotten. The result is injury or even death. One man had an electric staple gun jam. He used a needle-nose pliers to try removing the jammed staple while the unit was still plugged in. Simultaneously, he jerked out the bent staple, accidentally hit the on switch, and quickly drove five staples to clamp his shirt front to his belly button before he could yank the power cord loose. Worse accidents happen. You could loose a finger, limb, or your life by being uncautious with a power tool.

Jigsaws, circular saws, electric drills, sanders, grinders, and electric screwdrivers are the most common of household power tools. Then there are many more specialty items: electric lug wrenches, electric hammers, electric staple guns, electric pencil sharpeners, and electrically driven vises, vibrators, files, spray painters, tile glue applicators, cable cutters, and so on.

There are also miniature battery-operated versions of the larger power tools for hobbyists involved in model train and airplane building. For these, the only real differences are the size of the tools and the dc motor used.

Always study the tool first to see if a mechanical malfunction is involved. Often a motor cannot turn a shaft because of some sort of mechanical bending or binding. The tool has been dropped, or improper force has been applied in pressing the tool against some surface. A shaft has been bent, set screws may have come loose, or part of the appliance casing may be jamming the proper operation of the tool.

Rotate shafts by hand, with the unit disconnected from all power sources. Such visual inspection can reveal the trouble. Sometimes it will be necessary to remove the casing in instances where the mechanical jam is internal. Check to see that any gears are meshing correctly, that any belts, pulleys, or cams are rotating properly, and that the device is free of obstructions.

Frequently, power tool jam-ups are caused by scrap material getting inside the works somewhere. On a power saw or in an electric drill, scraps from previous jobs may have accidentally slipped inside the case and jammed the unit. In power tools without overload protection devices, the jams can lead to the burnout of the electric motor. Replacing the motor without fixing the jam will not repair the tool.

As in most small appliances, the most common problems arise from damaged power cords, wall plugs, and on/off switches. The repair or replacement of defective cords, plugs, and switches is generally a simple matter. Exact replacements are not necessary, so long as the new part you purchase is of suffient size and current-carrying capability to supply voltage to the tool. Replace a three-wire grounded cord for a power tool with another three-wire cord, *not* with an ungrounded two-wire connector. The same goes for plug replacement.

In switch replacement, be sure the new switch is of the same type. Don't substitute a single-pole/single-throw switch into a tool designed to use a double-pole/double-throw switch. If a substitute replacement switch does not fit inside the housing of the tool, don't just tape it in place somewhere on the handle with exposed wires. This can be a dangerous stop-gap repair. In fact, don't use that substitute at all. Find a suitable replacement that *will* fit.

You will frequently encounter sealed motor units on small power tools. If the motor casing is held together by rivets or welding, getting at the motor will be difficult or impossible. Since the motor is useless unless you *can* effect a repair, you might as well carefully pry or drill the case apart. If you fail, nothing is lost since you would have to purchase a replacement motor anyway. And you might find that the needed repair is simple and easy. The motor may be fine, but an internal contact wire is loose or corroded. Make the repair, and then use sheet metal screws or nuts and bolts and a drill to carefully put the case back together again. (Be sure that the screws don't stick out where they are not supposed to.)

The motor may have been damaged mechanically by a bent shaft, or by broken bearings or mounts, etc. If an exact replacement cannot be found, it is often possible to take the broken or bent part to a local machine tool shop where an operator can fashion a replacement less expensively than buying a new motor. Sometimes, the machinist will be able to straighten or repair the broken or bent part without weakening the structure excessively. There have been cases where a $1 bench straightening job, done in seconds, meant that the owner avoided having to purchase a new $150 replacement of the entire tool.

On power tools, study the unit to determine if some form of magnetic clutch, speed control, or overload device has been incorporated into the design. Often, the malfunction is due to one of these control units, rather than due to damage of the motor itself. Test and repair or replace these parts before giving up on the motor.

In belt or gear-driven equipment, test for slippage and wear. Dress belts with soap or beeswax before ordering replacement. With gear arrangements, slowly turn the mechanism by hand to see where the meshing problem occurs. Often a tightened set screw or a repositioning of internal adjustments can "fix" the tool. Be sure to study the owner's manual to determine the location and method for making possible tool

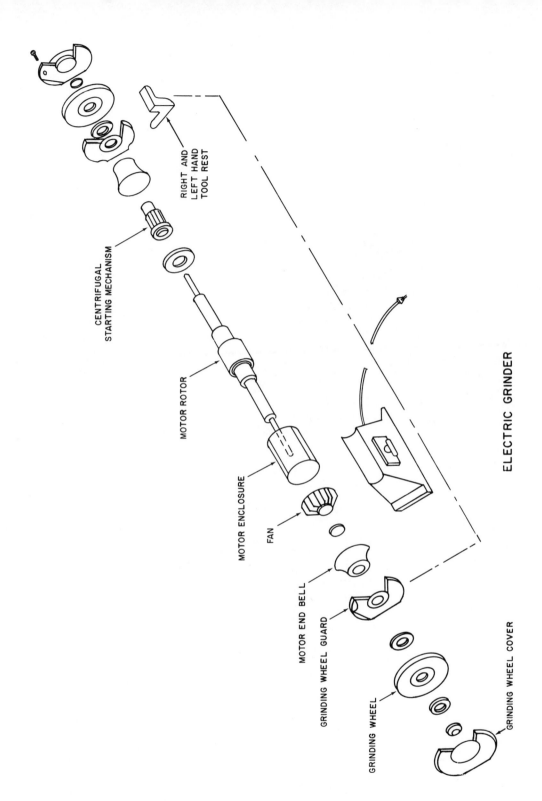

RIGHT AND
LEFT HAND
TOOL REST

CENTRIFUGAL
STARTING MECHANISM

MOTOR ROTOR

MOTOR ENCLOSURE

FAN

MOTOR END BELL

GRINDING WHEEL GUARD

GRINDING WHEEL

GRINDING WHEEL COVER

ELECTRIC GRINDER

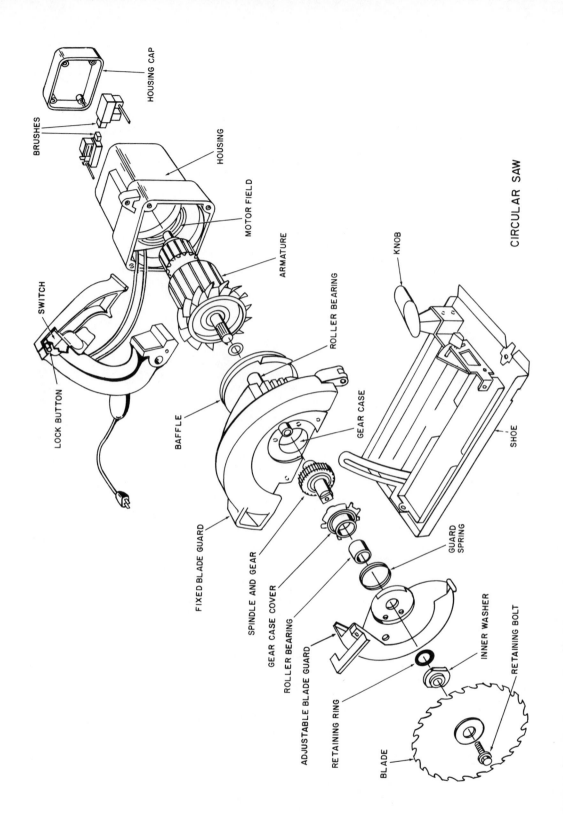

CIRCULAR SAW

HOUSING CAP

BRUSHES

HOUSING

MOTOR FIELD

ARMATURE

SWITCH

LOCK BUTTON

BAFFLE

ROLLER BEARING

GEAR CASE

FIXED BLADE GUARD

SPINDLE AND GEAR

GEAR CASE COVER

ROLLER BEARING

ADJUSTABLE BLADE GUARD

RETAINING RING

KNOB

SHOE

GUARD SPRING

INNER WASHER

RETAINING BOLT

BLADE

196

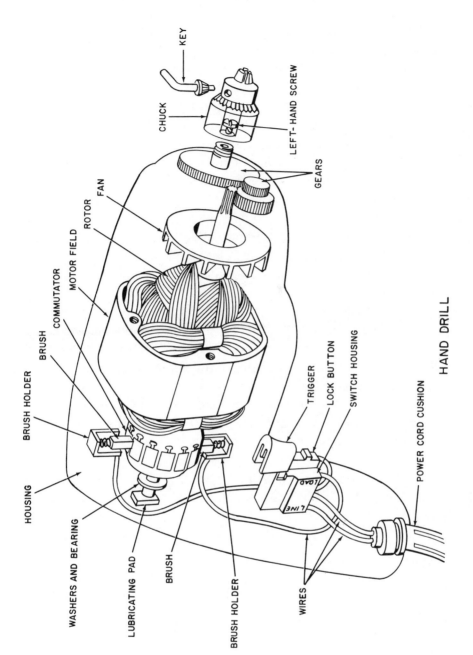

KEY

CHUCK

LEFT-HAND SCREW

GEARS

FAN

ROTOR

MOTOR FIELD

COMMUTATOR

BRUSH

BRUSH HOLDER

HOUSING

WASHERS AND BEARING

LUBRICATING PAD

BRUSH

BRUSH HOLDER

WIRES

TRIGGER

LOCK BUTTON

SWITCH HOUSING

LOAD

LINE

POWER CORD CUSHION

HAND DRILL

Reprinted by permission from *Reader's Digest Fix-It-Yourself Manual* (New York: Reader's Digest General Books, 1977).

adjustments. It may not be broken—just worn to a point where a standard operator's adjustment was supposed to have been made. And you forgot to review the manual carefully before using the tool extensively.

Impact tools and those involving some kind of back-and-forth movement for proper operation can be designed around cams, automatic voltage switching arrangements, and belt/pulley/gear drive mechanisms. A cam is an eliptically shaped gear that moves a part in and out as the cam shaft rotates. An example would be a cam wheel pressing against the drive shaft of a record player or tape recording turntable. When speed changes are made by the operator, the cam rotates and causes a different set of pulleys, belts, or drive gears to be engaged. The motor works at a constant speed, but a mechanical change effected by the rotating cam changes the speed of the device driven by the motor.

The cam is usually held in place by a set screw. It won't do the job if that set screw has come loose, or has fallen out of place.

Other impact tools utilize solenoids. This is a type of electric switch. A coil of wire is used as an electromagnet. When power is applied to the electromagnet, it will either pull in or push out a metal shaft mounted inside the core. The movement of the solenoid shaft will either mechanically turn some switch on or off, or will reposition some part of the tool's mechanism in order to switch its operational function. Solenoid coils may be checked with a VOM, just like testing the windings of a motor. Test for continuity in the resistance range (with no voltage applied). Then use the VOM to test all contacts and all wiring to and from the solenoid. With voltage off, push and pull the solenoid shaft to see that it moves freely and is not jammed. It can be lubricated, just like a motor shaft.

DISASSEMBLY

With any power tool, disassemble slowly and cautiously. Many tools contain springs and special clips that help to position critically placed parts. Careless disassembly can cause springs to fly across the room, lost forever—and you won't be sure of where a replacement goes. A lot of gears, pulleys, drive shafts, and mechanical parts are held in place with spring clips. Often these must be replaced when they are removed during disassembly, because the removal bends them slightly out of shape. Just be sure the fittings are as tight upon reassembly as they were when you first took the tool apart.

Again, the necessity of using the right tools cannot be emphasized too much. *Don't strip screw heads or nuts or bolts.* Use careful force in "breaking" any connection or tight screw or bolt. The use of WD-40 or some other type of penetrating oil can help in the removal of stubborn nuts and bolts.

When it comes to repairing broken wires or electrical contacts, review the soldering splicing tips in this book. Do the repair right, and the fixed electric power tool will give you years of smooth operation.

Troubleshooting Chart

Possible Causes	Solutions
Problem: Tool does not turn on; on lamp does not light.	
Faulty power cord or plug	Check cord and plug
Defective switch	Check switch
No power to outlet	Check outlet
Problem: Breaker or fuse blows when unit is plugged in or turned on.	
Circuit overload	Try another outlet supplied by a different circuit
Short circuit	Test
Problem: Unit turns on, but does not rotate properly, or tool does not do its work properly. Tool overheats.	
Blown fuse or breaker	Test
Motor winding is open	Test, replace
Connecting wires from power supply to motor are open	Test, replace
Thermostat is bad	Test, replace
Grime/dryness is keeping motor plate from working	Clean unit/lubricate
Mechanical binding	Examine tool; repair or replace as necessary.

Pumps

A pump sucks up liquids from one point and uses pressure to force the liquid through a pipe or tube to another location. Most pumps used in small appliances or in household operations are electrically powered. They include pumps in automatic dishwashers, sump pumps to drain flooded basements or patio areas, siphon pumps for moving gasoline from one storage tank to another, and pumps in an evaporative cooling system that keep the system's cooling pads moist.

Most electrical pumps receive operating current through a grounded three-wire cord because of the dangerous combination of electrical power and moisture. Part of household routine maintenance procedures should be a monthly check to test the electrical continuity of all pump ground connections. For shock safety, insulated plastic pump housings are better than metal housings.

Power cord insulation is extremely important for safe pump use. The electrical connections to all outdoor pump motors should be thoroughly checked at least every 60 days for fraying or weather wear.

The principles of pump operation are simple. An electric motor drives a shaft. The shaft extends into a pump suction head, that is attached to vanes or blade-like extensions. The suction head is immersed into the liquid to be sucked up. This suction head is in a covered housing, in which the vanes rotate. Centrifical force whirls the liquid through the suction head housing, which channels the flow into a pipe which then sends the liquid on its way. Suction heads are generally equipped with a strainer or filter so sediment or materials floating in the liquid can't get into the suction head and jam the rotating motor action.

With vacuum pumps, for lift-type suction, the vanes whirl in an airtight chamber channeling air from the chamber and the lift pipes through a vent. Thus centrifical force is used to generate a vacuum. The resultant vacuum is strong enough to pull the liquid up from its lower level to a point as high as twenty feet above the surface.

In addition to the normal on/off power switch, most pumps have a float valve that works as a secondary power switch. The float is a hollow plastic or metal cannister, or a wooden or cork float. It is attached by lever to a mechanical power switch. The lever may be adjusted so that the motor turns on or off when the liquid reaches a desired level. The float is much like the big floating bulb in a toilet tank, which turns off the water input when the tank is filled. In the case of a pump, the float acts to turn the motor on or off at certain preset levels.

Complex pumps also incorporate check valves, or vacuum valves, which allow liquid to flow through them in one direction, but be blocked from backing up in a reverse flow.

Before attempting any extensive pump repair, be sure to check the system thoroughly for clogs and mechanical jams. These are the main reasons a pump fails to operate correctly. Also, the suction head must be fully immersed or the pump will not operate.

If a check valve is part of the system, this may become clogged and defective. A hinge action inside the valve allows a plate to move back and forth, opening to allow flow in one direction, and swinging back to close the pipe when the liquid attempts to backflow. A check valve can become inoperative if the hinge or the plate is jammed by sediment; better systems use a vacuum valve that is self-cleaning.

DISASSEMBLY

Since many parts of a pump operate under water or while immersed in some possibly corrosive liquid, extreme care should be taken when attempting to take down the unit for service. Rust or corrosion may literally weld nuts onto bolts, and screws into "permanent" sockets. Use liberal application of a penetrating oil such as WD-40, and try to avoid excessive force in "breaking" stubborn connections. The same rust and corrosion problems might make it difficult to take apart pipe and valve connections.

The pump motor itself will be housed away from the liquid. As a general rule, a shaft will lead from the motor through a watertight gasket and to the impellor of the pump. If this gasket fails and fluid gets inside the motor housing, you'll find yourself replacing the pump — or at least the motor — rather soon.

The housing is often held together with long bolts. With some pumps, the housing is a sealed unit. This helps to make the motor last longer, but it also means that you'll have to replace the entire pump in case of a motor problem.

Wear and rust mean that floats and their lever attachments, and many components of the suction head, might have to be replaced fairly often. A regular maintenance schedule, with complete cleaning and lubrication, will help everything last longer.

Many of these related parts are held in place by screws, clips, or other devices. Once again, rust and corrosion are the greatest enemies. Removal for disassembly can be a frustrating job. Do so carefully.

Troubleshooting Chart

Possible Causes	Solution
Problem: Pump does not turn on.	
Faulty power cord or plug	Check cord and plug
Defective switch	Check switch
No power to outlet	Check outlet
Problem: Breaker or fuse blows when unit is plugged in or turned on.	
Circuit overload	Try another outlet supplied by a different circuit
Short circuit	Test
Problem: Unit turns on, but does not rotate properly or pump liquid as designed. Pump overheats.	
Blown fuse or breaker	Test
Motor winding is open	Test, replace
Connecting wires from power supply to motor are open	Test
Float valve is stuck	Test, replace
Grime or mechanical jam	Clean unit/lubricate

Sewing Machines

Before attempting to repair a sewing machine study thoroughly the operator's manual. If the manual has been lost, obtain a replacement from the manufacturer before starting your repair efforts.

When purchasing a new sewing machine, carefully review the normal maintenance procedures for that particular model. Frequently, there are points that must be lubricated *before each use* of the machine. Also, there may be switch and lever positions to be set when the machine is not in use for long periods of time. The threading and needle change information is critical, since needles break frequently, and the machine will not operate if threaded improperly.

Proper oiling is a must. When in continuous use, the machine should be oiled daily. Oiling spots are generally inside the arm and in the bed of the machine (see the owner's manual). Generally, the bed is accessed by lifting the hinged sewing machine plate and swinging it back. Apply a drop of oil to all moving parts.

Many machines have sealed motors that call for no lubrication. However, a drop of oil on the shaft as it enters the motor housing can be helpful.

The shuttle race is one of the most important oiling spots on the sewing machine. A drop of oil in the spot indicated in your owner's manual should be applied before each use of the unit.

There should be no more than $1/4$ to $1/2$ inch of slack in the motor belt. If excessive, locate the adjustment screw, which is usually located on the motor holding bracket. Loosen this screw, retightening it when the belt is back to its proper tension. Often there are alignment adjustments for the drive belt. Consult the manual for your particular model.

In many models, bobbin thread tension is permanently set at the factory and seldom needs changing in normal operation. But when the tension of the bobbin thread is consistently wrong, you can adjust it by turning the screw on the bobbin tension spring. To increase tension, turn the screw slightly to the right. To decrease tension, turn the screw to the left.

For a jammed shuttle, first clean away loose threads and debris. Turn the balance wheel to lift the needle to its highest position, then remove the bobbin case. Press down on the spring-loaded shuttle race cover clamps and remove the shuttle race cover and shuttle body. Carefully clean by removing all stray thread, lint, etc., and reassemble.

Missing stitches are a common problem, but they are usually due to operator, not machine, error. If the machine is not threaded correctly, the machine can jam up, skip stitches, stitch irregularly, and any number of other problems.

Improper stitching and thread breaks can also be caused by improper adjustments. A standard sewing machine has a number of adjusting screws and knobs and so on. Set the bobbin tension too high, for example, and that thread is bound to break. (Set it too low and the machine won't sew correctly.)

Thread breaks can also be caused by using the wrong needle, or by using a needle that is bent or damaged in some way. A needle of the wrong length can also break, or can cause other damage to the sewing machine.

If your sewing machine is having a problem with stitching, and you just can't seem to find the source of the difficulty, try loading the machine with two contrasting thread colors, such as black and white. Then look at the results. Quite often you'll be able to tell quickly if the trouble is with the upper or lower sections of the threading mechanisms.

Sometimes when a sewing machine is brought in for repairs, the owner says that everything seems to work okay, but the machine runs "heavily." The first and immediate suspect is dirt, grime, lint, pieces of thread, and other such things. Most of the time, a decent cleaning is all that is needed to get the machine back into tip-top condition again.

The use of inferior types of lubricating oil can also cause problems and sluggishness, as can over oiling. (In correct lubrication, a little goes a long way.) Follow the manufacturer's recommendations. For lubrication, tension settings, other adjustments, and disassembly your best source is the owner's manual. An owner's manual is important for all appliances. With a sewing machine it is absolutely essential.

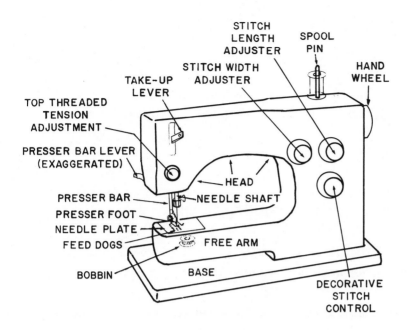

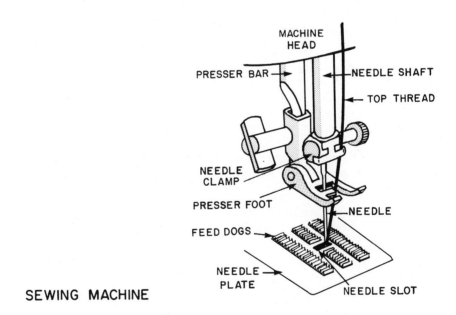

SEWING MACHINE

Reprinted by permission from *The Complete Book of Machine Embroidery* by Robbie and Tony Fanning (Radnor, Pa.: Chilton, 1986).

DISASSEMBLY

Each brand and model is slightly different. More important, due to the precision of the machine, you probably should not try to further disassemble your sewing machine, other than what is detailed in the owner's manual. This is one time when leaving the job to a professional is worth the expense.

Troubleshooting Chart

Possible Causes	Solution
Problem: Sewing machine does not function; on lamp does not light.	
No power to outlet	Check outlet
Faulty power cord or plug	Check cord and plug
Defective switch	Check switch
Problem: Breaker or fuse blows when unit is plugged in or turned on.	
Circuit overload	Try on another outlet supplied by a different circuit
Short circuit	Test
Problem: Unit turns on, but machine does not sew properly.	
Connecting wires from power supply to motor are open	Test
Grime/lint in motor	Clean unit; lubricate
Drive belt broken or slipping	Examine; repair or replace
Problem: Threads (upper and lower) break.	
Incorrectly threaded	Rethread
Tension too tight	Adjust
Machine dirty and clogged	Clean; lubricate

Troubleshooting Chart

Possible Causes	Solution
Problem: Fabric does not move through properly.	
Improper tension or other settings	See owner's manual and adjust
Worn or damaged feed dogs	Examine; repair or replace
Problem: Skips stitches, or gives irregular stitches.	
Improper threading	Rethread machine
Tensions not correct	Adjust
Bad or improper needle	Replace
Dirty or clogged mechanisms	Clean

Shavers

Cleaning, maintenance, and repair of electric shavers depends much upon the model, so refer to the owner's manual first. Each model incorporates differing cleaning and lubricating methods, and although most razors have a cutting head that simply pops off for cleaning, disassembly can be tricky if you don't have the proper manual for your razor. If you have misplaced the manual, generally a retail store which sells the same make and model will be happy to provide you with a replacement manual at low cost. Or you can write directly to the manufacturer.

Rechargeable cordless shavers are very popular today. Some have a wall plug that can be left in an outlet for a few hours until the razor's internal battery recharges. Others are entirely cordless, and the holding stand is the recharging device. Again, reference to the operator's booklet is essential. Some units should be left plugged into the wall outlet continuously, while others can be damaged if overcharged. A few models operate either from a standard ac wall outlet in the home, *or* from the cigar lighter attachment in a car (a dc voltage source). Some shavers have a strong enough internal rechargeable battery that they can be used for five to ten days before running down. (These are great for camping trips.)

Cordless shavers have inside battery packs. These can be tested in the same way as any other battery or battery pack. Use your VOM, set to read dc volts in the proper range (usually well marked on the battery).

The battery pack itself might be the slip-in type, which makes changing it relatively simple. All too many models use a pack that solders into place. Changing these can be tricky, and even dangerous. The rechargeable batteries are sensitive to heat, and it's easy to damage the cells during soldering. And with the cost of the replacement pack, you might want to let a professional handle the battery change rather than taking the chance of ruining the new pack.

Before going through the bother and expense of changing the battery pack, take a moment to test the output of the charging base unit with your VOM. If it's not supplying the proper voltage, the batteries in the shaver won't get recharged. What appears to be a problem with the batteries could have nothing to do with the shaver itself, but with the recharging unit.

Check with the owner's manual for your particular shaver to see if there are adjustments for various types of hair—thick, thin, dry, oily. Also test to see which operating positions work best for you. Normally, the shaver head should be kept flat against the surface to be shaved. Use short and even strokes, up and down about 1 to 2 inches. Do not move shaver in a sideways or rotary direction, and always move the head with and against the direction of hair growth.

Most shavers use one of two types of motor. Perhaps most common is the simple universal motor (see Chapter 5). This is like a miniature version of any other standard universal motor. The same principles apply, as do testing and repair techniques.

Also common is the vibrator motor, which isn't really a motor at all, but more like the clapper used to ring the bell in an old-style telephone. The vibrator causes a back-and-forth motion due to an electromagnet pulling on a spring which is in turn attached to the cutting blades. Testing it is much the same as testing a standard motor. If power isn't getting to the electromagnet, nothing will happen, so your first step is to use the VOM to test for incoming power. The electromagnet is a coil of wire wrapped around a metal core. As with the windings of a motor, the winding of an electromagnet can be tested for continuity, for short circuits, and for opens.

Whatever drive system is used, always begin with a thorough cleaning and an equally thorough visual examination. More times than not, the problem will be simply one of hairs causing a jam or other malfunction.

DISASSEMBLY

The quality of shave you get will be consistently good if you regularly disassemble and clean the head after daily use. Routine lubrication with a light shaver oil will keep the cutter blades sharp and the movement swift and sure. The owner's manual will explain how to do this properly.

Almost all shavers and clippers use very small motors. These are

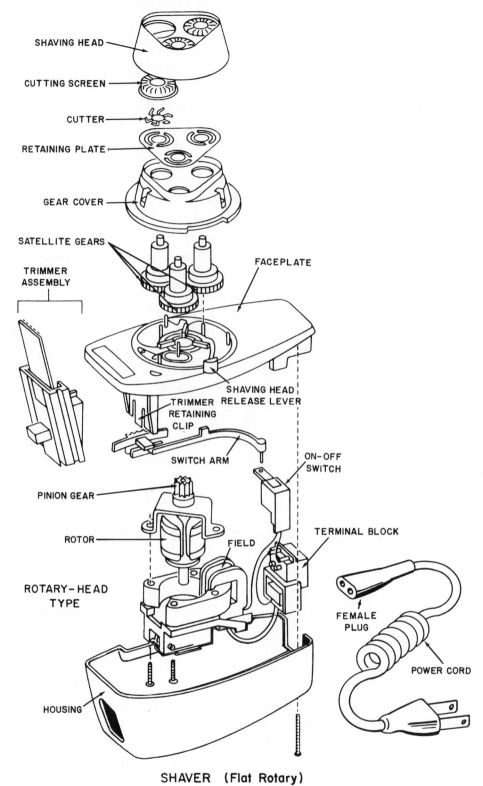

SHAVING HEAD

CUTTING SCREEN

CUTTER

RETAINING PLATE

GEAR COVER

SATELLITE GEARS

TRIMMER ASSEMBLY

FACEPLATE

SHAVING HEAD RELEASE LEVER

TRIMMER RETAINING CLIP

SWITCH ARM

ON-OFF SWITCH

PINION GEAR

ROTOR

FIELD

TERMINAL BLOCK

ROTARY-HEAD TYPE

FEMALE PLUG

POWER CORD

HOUSING

SHAVER (Flat Rotary)

Reprinted by permission from *Reader's Digest Fix-It-Yourself Manual* (New York: Reader's Digest General Books, 1977).

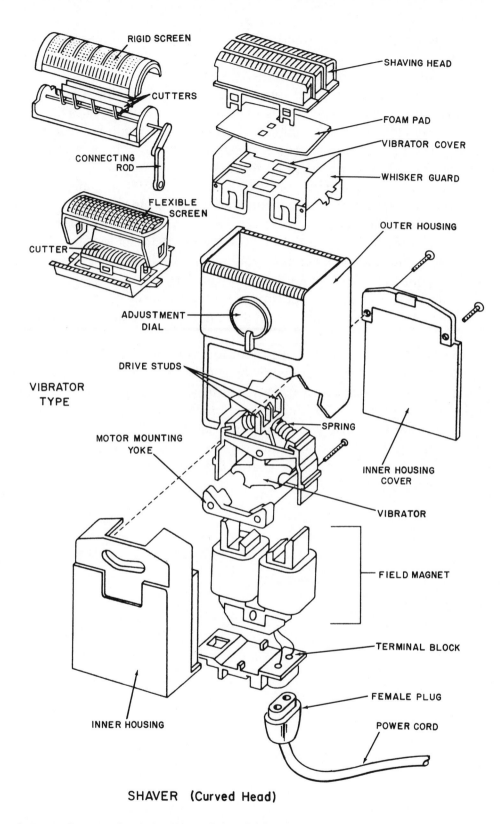

RIGID SCREEN

CUTTERS

CONNECTING ROD

FLEXIBLE SCREEN

CUTTER

SHAVING HEAD

FOAM PAD

VIBRATOR COVER

WHISKER GUARD

OUTER HOUSING

ADJUSTMENT DIAL

VIBRATOR TYPE

DRIVE STUDS

MOTOR MOUNTING YOKE

SPRING

INNER HOUSING COVER

VIBRATOR

FIELD MAGNET

TERMINAL BLOCK

FEMALE PLUG

POWER CORD

INNER HOUSING

SHAVER (Curved Head)

difficult to repair, and are sometimes almost as difficult to remove for replacement. The screws and bolts tend to be rather tiny, and they sometimes require special small-bit tools.

Most modern razors have some style of snap-apart disassembly for the case, with access screws involved only when you must get to the motor or the power switch assembly. If you do not have a manual, carefully study your unit to see if you can determine exactly how the housing goes together and what holds the head to the top of the appliance.

With the head assembly open, carefully lift any moveable hairstoppers, oscillators, and cutting blades. Use a cleaning brush, usually supplied with the razor, to carefully remove any hair clippings.

Many manufacturers provide convenient customer service. Every six to twelve months, you can take your shaver to the store where it was purchased (or to a regular manufacturer's outlet), and a trained operator will clean, oil, and adjust the cutting blades. This service is free in many instances, and available at a minimal charge for other makes and models.

In case of a total power supply malfunction (the razor does not operate, blows fuses or breakers when plugged in, does not recharge, etc.), then it will be necessary to break down either the shaver itself or its companion recharging unit. Although the motors are tiny, the same service procedures are involved as outlined in Chapter 5. The workshop VOM may be utilized to determine cord, plug, and switch defects; the voltage and current scales are useful for testing the recharging capability of any plug-in attachments.

Troubleshooting Chart

Possible Causes	Solution
Problem: Shaver does not operate.	
No power to outlet	Check outlet
Faulty power cord or plug	Check cord and plug
Defective switch	Check switch
Recharger faulty	Check charging unit
Faulty motor	Test; repair or replace
Problem: Breaker or fuse blows when unit is plugged in or turned on.	
Short circuit	Test
Problem: Unit turns on, but does not shave properly, or recharge as designed.	
Connecting wires from power supply to motor are open	Test
Charger is faulty	Test/replace
Grime/mechanical jam of cutters	Clean unit/lubricate
Transmission is broken	Disassemble and examine
Cutting blades dull or damaged	Examine; replace

Skillets and Frypans

One of the first small appliances to find its way into numerous homes was the electric skillet. Today these electric cookers come in just about every conceivable size and style. There are tiny frypans to handle a couple of strips of bacon and a single egg, deep fat fryers for whipping up a batch of donuts or french fries, huge "serving tray" cooker/warmers for handling a half ton of hors d'oeuvres, special pans to shape and cook single hamburger patties, and on and on.

A development which has added immensely to the popularity of these items is that most of the models manufactured over the past decade have no electronic parts except the heating element contained in the under surface of the pan. All power switches, thermostat elements, and connecting cords are self-contained in a removeable handle. This means that the dirty skillet can unplugged and fully immersed in soapy water for cleaning. Before you do so, however, read the instruction manual that came with the appliance. It will tell you how to properly clean the electric fry pan.

Despite the great variety in design, most of these appliances work on the same general principles and have very similar parts (and very common repair problems).

A fairly common problem with a pan that can be washed and a plug-in control unit handle is corrosion of the plug-in contact points, due to frequent wetting. One way to keep these electrical contacts clean and in good conducting shape is to use a fine sandpaper or a fine metal file (*gently!*) about once a month to lightly touch up the contact points. Be sure to thoroughly blow out any metal filings.

Another way to clean contacts that leaves no residue is to take a *non-lubricating* spray contact cleaner and treat the two plug surfaces liberally once every four to six weeks. Spray the prongs on the handle, and spray into the prong socket; insert the plug and wriggle it around, then remove.

A disadvantage of the immersible style frypan is that usually the heating element is sealed inside a cavity in the base of the pan. If something happens to cause the heating element to open (fail) the pan must be returned to the manufacturer or a company service representative for repair or replacement. Quite often a shorted or opened heating element means that it is time to replace the appliance as a whole. Study your unit carefully to determine if the heating element is accessible for service, or whether the pan and heater are fully sealed.

DISASSEMBLY

The control handle for electric skillets are usually very easy to take apart, and service checks are quite simple. In most cases, no more than one or two bolts or tap screws are used to hold the handle together, or to attach a cover plate for the power controls. Many of these detachable control handles are of the "clam shell" type—two halves held together with a single nut and bolt protruding through the middle. When taken apart, one half of the shell is the cover, and the working controls are held in place in the other half.

The VOM may be used for continuity and voltage checks in testing the power cord and plug, the on/off switch, the thermostat or rheostat temperature controls, and any internal timing controls the appliance may utilize.

In some deep fry models and in many of the older electric frypans, the control units are contained with the heating element in the bottom of the pan. For these, disassembly is generally possible by turning the appliance upside down on the workbench in order to gain access to the hardware holding the unit together. Check the heating element and any thermostat or rheostat temperature control according to the tips given in Chapter 6.

Replacement parts will be obtainable from the manufacturer (consult the operator's manual) or the nearest manufacturer's service representative. For some standard brand electric skillets, common replacement parts might be available at major discount supermarkets or appliance warehouse outlets.

Defective power switches, cords, and plugs may often be replaced by parts available from any general hardware counter, so long as the replacements are of the same wire gauge and switch type, and can fit the available space.

Troubleshooting Chart

Possible Causes	Solution
Problem: Device begins to overheat; unit does not turn on; pilot lamp does not light.	
Faulty power cord or plug	Check cord and plug
Defective switch	Check switch
No power to outlet	Check outlet
Problem: Breaker or fuse blows when unit is plugged in or turned on.	
Blown fuse or breaker	Test
Heat element is open	Test/replace
Connecting wires from power supply are open	Test
Thermostat is bad	Test/replace
Grime or dryness is keeping unit from working	Clean unit

Slow Cookers

Slow cookers are similar in operation to electric skillets. The difference is in the thermostat that determines the amount of heat being produced and in the electro/mechanical timing devices.

Some of these units are designed to operate with a clamped and sealed lid, much like pressure cookers. In such instances, the appliance is equipped with a steam escape valve and generally with some form of sensing device to automatically turn off the cooker if the food becomes too dry (to avoid burning). If the valve or sensor fails to operate properly, the owner can come back to find a sticky, burned mess.

Many of these units are made of ceramic, which is easily cracked or flawed. Some have containers with double-wall construction (an air space between the walls). Leaks, cracks, and material flaws can make the unit irreparable. Some of the better units, however, can be supplied with replacement pots and lids so long as the heating element, thermostat unit, and all other electrical parts are in good shape. Study the operational manual to determine what replacement parts might be available in case of malfunction.

Power cords, plugs, and on/off switches cause most of the breakdowns of this appliance, with defective heating elements or thermostats coming in second. Testing and repair of cords, plugs, and switches is reviewed thoroughly in Chapters 3 and 4.

Slow cookers are intended to be in place and on for very long periods of time while unattended. For this reason, the appliance should be carefully placed for safety when operated unattended. Make sure no flammable materials are close, in case of a short or other electrical malfunction. And be sure the unit is placed on some type of rack or raised surface, so that in case of overheating or cracking, fluids will not become

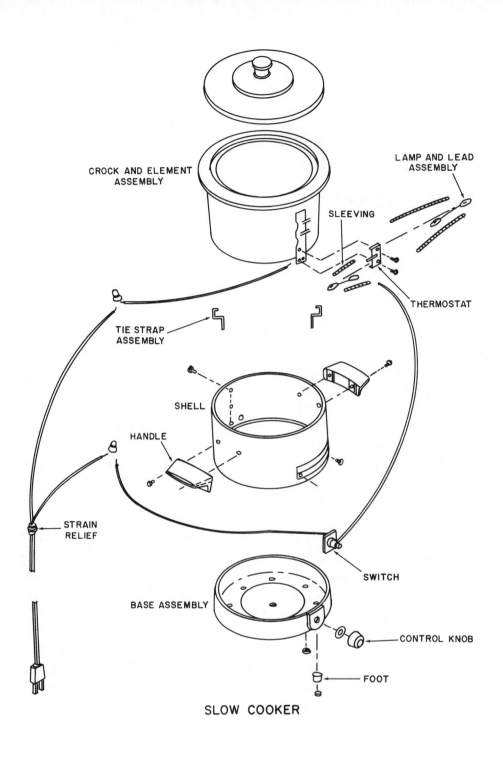

CROCK AND ELEMENT
ASSEMBLY

LAMP AND LEAD
ASSEMBLY

SLEEVING

THERMOSTAT

TIE STRAP
ASSEMBLY

SHELL

HANDLE

STRAIN
RELIEF

SWITCH

BASE ASSEMBLY

CONTROL KNOB

FOOT

SLOW COOKER

a danger if they overflow and seep into the heating element or electrical control circuits. Many safety-conscious cooks will operate these appliances only on drainboards where there is a slight downgrade flow to a sink, or outdoors where malfunctions cannot lead to problems beyond the cooker itself.

Follow cooking instructions that come with the appliance. Do not overfill, and watch the temperature and time control settings carefully.

Cleanliness and maintenance are important to obtain long life. Follow the manufacturer's cleaning instructions, and do not immerse the heating element or the electrical circuits in water.

DISASSEMBLY

A few of the least expensive slow cookers have electronic components that are completely sealed inside the ceramic cooking pot. These are not possible to repair, since opening the unit will crack the cooking container.

Appliances that may be repaired generally come apart from the bottom. Empty and clean the cooker according to the operating manual, and turn the pot upside down. Removal of two to six screws or holding bolts will usually allow removal of the bottom and provide access to the power switch, control devices, and heating element.

Use your VOM to check for continuity and to test voltages at various operating points. This, plus a careful visual inspection, will allow you to find most malfunctions. Because of tight spacing and the critical cooking temperatures required for proper operation, exact replacement parts will be necessary in most instances.

In case of container leaks, check any rubber or plastic seals for wear or breakage. Some commercial ceramic glues and repair solvents are available, but since the appliance will be used under heat, use only fire-retardant adhesives in attempting repairs. And check to be sure that any glue or solvent used does not deteriorate under heat, where it might become dangerously absorbed into any food being prepared in the appliance.

Troubleshooting Chart

Possible Causes	Solution

Problem: Device begins to overheat; unit does not turn on; pilot lamp does not light.

Faulty power cord or plug	Check cord and plug
Defective switch	Check switch
No power to outlet	Check outlet

Problem: Breaker or fuse blows when unit is plugged in or turned on.

Circuit overload	Try on another outlet supplied by a different circuit
Short circuit	Test

Problem: Unit turns on, but does not cook properly.

Blown fuse or breaker	Test
Motor winding or heat element is open	Test/replace
Connecting wires from power supply are open	Test
Thermostat is bad	Test/replace
Grime or dryness is keeping unit from working	Clean unit/lubricate

Space Heaters

Portable electric heaters are a handy and relatively inexpensive way to warm small sections of a home. They are frequently used in bathrooms or bedrooms to keep things warm while bathing or dressing, or in a garage. Most space heaters are very simple in design and operation. They consist of a power cord, the cabinet, a heating element, the on/off switch and, in some better models, thermostat or rheostat controls to maintain certain temperatures. A few have motor-driven fans to help disperse heat throughout a small space.

As with any appliance that heats, electric space heaters draw a large amount of current. All replacement cords and parts must be of sufficient size to handle that current.

If your space heater seems to be constantly blowing fuses or circuit breakers, it could be that there is a short somewhere in the unit. More likely, the appliance is drawing more current than can be supplied by that circuit. The standard outlet in a home is supplied through a 15-amp circuit breaker or fuse. Many space heaters draw 20-amps or more. If you plug it in to a 15-amp circuit, it could seem that the appliance is malfunctioning.

Almost all space heaters will draw a respectable amount of current, even if it's less than 20 amps, and you can still have difficulties. For example, if the unit you purchase requires 14 amps to operate, if you also flip on a couple of lights, you'll have overloaded that circuit.

Most heaters are built with the electrical circuits all in series, so if any part of the circuit is open, the appliance will fail to operate. A total continuity check through the unit with the power off may help to pinpoint the problem. Begin by testing the overall unit (by testing at the plug), and then work your way inwards, isolating by sections, and then down to the individual components.

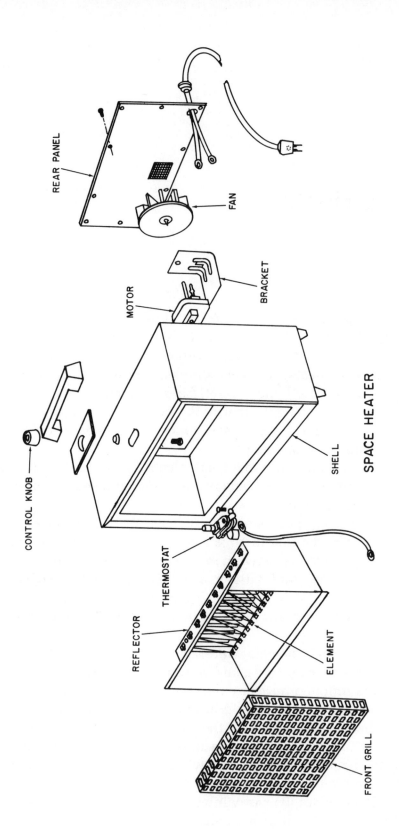

REAR PANEL

FAN

MOTOR

BRACKET

CONTROL KNOB

SHELL

THERMOSTAT

REFLECTOR

ELEMENT

FRONT GRILL

SPACE HEATER

222

The heater element(s) is held in place by insulated stand-offs. In most units the heater element is plugged. This makes it easy to test the element. Set your VOM to read resistance in the x1 range. A heater element in proper working condition will usually provide a reading of somewhere between 10 and 15 ohms. A reading of no resistance probably means that you're not taking the measurement correctly, but could also mean that there is a dead short in the element. A reading of infinity means that there is a break (an open) in the element and you'll have to replace it.

Occasionally you can replace just the element wire. This is done by feeding the new wire through the insulators. It takes a delicate touch. For the unit to work properly, the wire must be uniform throughout its path, and must be undamaged.

Other problems with space heaters include units that fail to go on at all, that short circuit, that do not heat even though the fan operates, that refuse to maintain the temperature at the selected control settings, or that give a shock to the operator when the cabinet it touched.

A visual examination of the appliance and its components will often reveal the source of the problem. Look especially for signs of burning and for obvious breaks in the element wire.

DISASSEMBLY

A space heater usually consists of the rear case, two side panels, the heater element and reflector assembly, and the front panel with the controls and the protective screen. All models will also have legs and a handle. It's rare that you'll ever have to remove these.

If the unit has a fan, it is probably attached by screws to the rear casing. This means that you have to be careful during disassembly. The screws that hold the case together are virtually always around the edges. If there are screws that are away from the edges, chances are good that they are holding something inside. Don't remove them until you know where they go and have determined that there is a reason for removing them.

Controls and other components on the front panel—such as the various switches and thermostat—are wired to the incoming power at the back panel. Often the leads will have plug-type connectors. Other times they will be soldered or otherwise hard-wired into place. Be very careful when separating the case, because you could easily damage the wires or the components to which they are attached.

During disassembly, you should be able to separate the case far enough to probe for continuity with your VOM. If you can't and have to unplug any wires, be sure to make notes and sketches so you know exactly how and where to reconnect the wires.

Some units also have tension springs, especially to keep the heater element tight.

Troubleshooting Chart

Possible Causes	Solution
Problem: Unit does not go on.	
No power	Check outlet
Circuit overload	Try another outlet, particularly one of at least a 20-amp rating
Bad cord	Test/replace
Bad switch or thermostat	Test/replace
Bad element	Test/replace
Problem: No heat, even though fan turns and cool air is blown out through the vents.	
Bad element	Test/replace
Bad thermostat	Test/replace
Problem: Temperature control is erratic.	
Bad thermostat	Test/replace
Problem: Heater causes a short circuit to the wall outlet when plugged in, even when on/off switch is in off position.	
Short in the incoming power wire	Repair immediately
Short in power switch	Repair immediately
Problem: Operator receives a shock when the heater cabinet is touched.	
Short in wire	Repair immediately
Problem: Heater works, but fan does not go on and air does not circulate properly.	
Bad fan	Test/replace
Bad fan switch	Test/replace

Telephones

With the changes in Ma Bell government regulations early in this decade, hundreds of small and large companies have begun to manufacture telephones for private use. It is now possible, and permissible, for homeowners to service their own telephone equipment when they have the necessary tools and knowledge. (If your repairs cause a problem with the phone lines, however, you are held responsible.)

For the most part, the tools you will need are the same involved in other small appliance repair and maintenance work. Most telephone repairs can now be made using no more than a little common sense in tracking down the defective part(s).

Many replacement parts are now available from general hardware and appliance stores that sells private telephone equipment. The most complex repair chores will involve phone ancillary equipment, such as telephone answering devices and recorders, or telephone amplifiers designed for "no hands" conversations or as aids for the hard of hearing.

You may also want to expand the household phone system. Consult a store which deals in the new phone equipment. There are jacks, plugs, wire, pre-made extensions, message recorders, and adaptors available for almost every conceivable need. Your local representative from the telephone company will be happy to advise you on what can and cannot be connected to existing lines and equipment that you now lease from the phone company.

When problems do arise and repair is needed, it's very likely that you can quickly and accurately diagnose the source of the trouble. The problem is either in the telephone or in the phone line. If other phones in your home are working, then you know that the trouble is either in that one particular outlet or in the telephone.

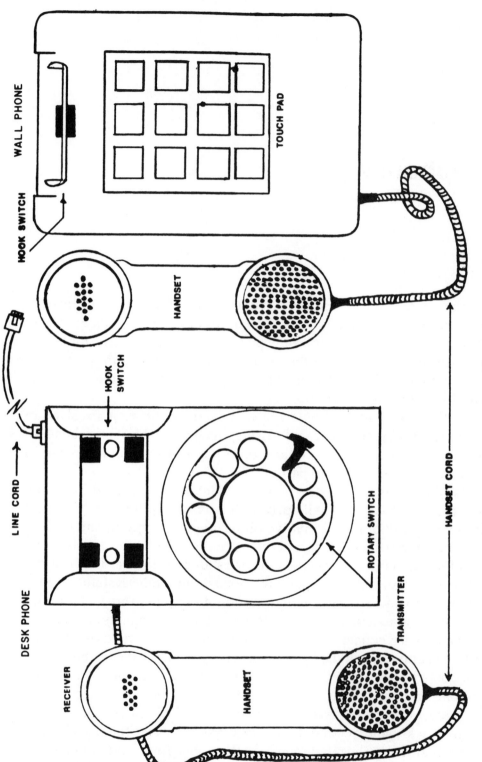

TELEPHONES (External Components)

Reprinted by permission from *Chilton's Guide to Telephone Installation and Repair* by John T. Martin (Radnor, Pa.: Chilton, 1985).

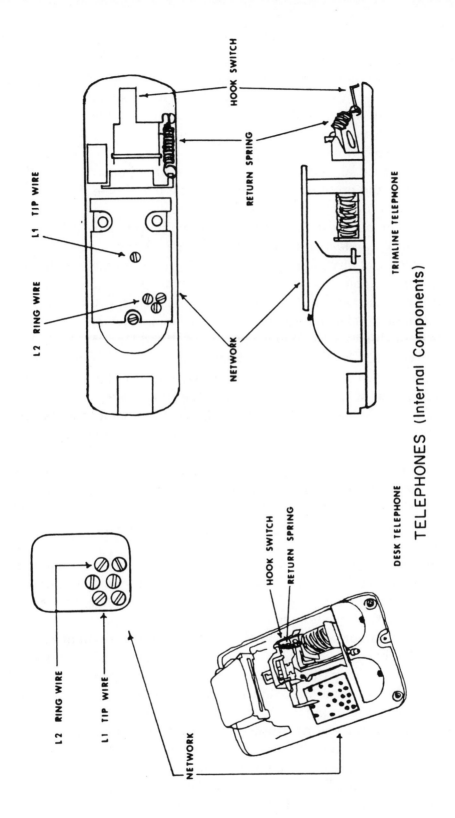

TELEPHONES (Internal Components)

HOOK SWITCH

RETURN SPRING

L1 TIP WIRE

L2 RING WIRE

NETWORK

TRIMLINE TELEPHONE

DESK TELEPHONE

HOOK SWITCH

RETURN SPRING

NETWORK

L2 RING WIRE

L1 TIP WIRE

227

Don't assume that a dead line everywhere in the house means that the trouble is with the incoming line. It's possible that a malfunction in one of the telephones or in one of the outlets is causing the entire house to go dead. One by one, unplug each phone, and listen to the receiver. If you suddenly hear a dial tone, you'll know where the problem is. It's with the last phone you unplugged, or with the last outlet from which you unplugged it.

Unplug the suspected phone and try another in that same outlet. If it now works, the problem is definitely in that telephone. If it still doesn't work, the trouble is probably with the outlet.

You can also test the suspected phone by plugging it into another outlet in the home or into an outlet at a neighbor's.

Disconnect any phone you are working on from the telephone company wiring. There is no danger of high voltage (although the voltage used to ring the bell can give you a shock), but when working on a phone you generally take the receiver "off the hook," which disables other units in the home. Also, when a receiver is lifted for over a minute without anyone dialing, the automatic equipment at the telephone company disrupts the dial tone and your phones will be dead for a while even after you repair whatever was wrong in the first place.

The phone receiver consists of the dial or touch-tone circuitry, a small carbon microphone, a tiny speaker mounted in the earpiece, a bell or buzzer, and the connecting wiring.

The receiver itself is so simple in design and construction that some small inexpensive phones are constructed in totally sealed plastic cases (many of them cost less than $10). No sense in breaking one apart for repair if it goes bad. It is cheaper just to purchase a complete replacement phone.

In better-quality phones, replacement hardware is available. If the user breaks one of the push buttons on the touch-tone board, or some of the other component, you might be able to buy just that part or section, and for less than a whole new telephone.

The major problem associated with phone equipment comes not from defective parts, but from bad or intermittant electrical connections. So before tearing apart a phone to see what might be the matter with it, plug another unit into the same jack outlet. Use spray cleaner on all the jack contacts, including those leading into the base of the phone or from the phone base to the mouthpiece. Check all cords and extensions for fraying or breaks, since it is easy for these light cords to be damaged during everyday use. A VOM can be used to test any cord or cable for continuity, which will tell you quickly if something has gone wrong with the fine wires inside that cord.

If all the phones in the home are of the same type and model, it is easy to track down a malfunction. Take a good phone and the bad one and, part by part, begin exchanging from one to the other. Switch microphones, receiver speakers, entire receivers from the base, touch tone boards, etc. When you find a part that works in neither phone, that is the replacement you will need to purchase. (*Note: Do not* disassemble a phone are leasing.)

Often you will find that every part from the bad phone works fine in the good phone! If, when you put them back into the bad phone, they still don't work, then you have a contact problem. Something is not making a good electrical connection. In many instances, you will find after testing the parts and then putting them back into the "bad" phone after you find they work in the good phone, that both units work. The switching around of components has cleaned up bad points on some of the contacts and you have "fixed" the trouble.

Amplifier phones have an additional component—an electronic amplifier to "build up" the strength of the voices coming into the receiver. This can be either a no-hands speaker phone for conference calls, or a hand set with a variable volume control to boost the signal for the hard of hearing.

If an amplifier goes bad, you can use the VOM to see if it is receiving its proper operating voltage. In some units, there is no power supply. The leased phone lines themselves provide an input of 9 volts dc from the phone company (to operate the bell and/or the touch-tone board). Part of this 9 vdc supply from the phone company is used to operate the amplifier. If the amplifier fails, check to see that the bell or buzzer and the touch tone board still work. You may have lost the dc voltage input from the phone company, and the problem is with the leased lines and not with your own equipment.

If you are familiar enough with transistors to be able to identify the emitter, base, and collector on individual devices, and if you know the proper voltage drop across a working transistor, then you can use the VOM to test which of the transistors may have gone bad. If not, or if the trouble is in an IC chip, it is usually best to replace the whole amplifier board. The cost of such replacement is usually no more than $15, and generally quite a bit less. Again, check the quality of those electrical contacts before purchasing a replacement.

ANSWERING MACHINES

With answering devices and their accompanying tape recorders, a worn or bad recording tape is a common cause of malfunctions. Check the owner's manual for proper head cleaning advice. Tape heads may be cleaned with either a commercial head-cleaning solvent from an electrical supply house, or with cottom swabs moistened with denatured alcohol. Do not use ordinary rubbing alcohol.

If the recorder has some sort of mechanical failure, disassemble the unit and use the information in Chapter 4 on electric motors to study the tape drive mechanism. Contact and defective cord or plug problems are more likely to occur than breakdowns in the solid-state electronic components, so use the VOM for continuity tests on all cords, plugs, and phone connectors.

LEASED LINE PROBLEMS

Occasionally, a hum or line noise will become highly distracting to phone conversations. If switching phones does not clean up the noise, it is possible the trouble is in the lines from the phone company. This is particularly the case in periods of damp or wet weather, where the moisture may be causing "ground leaks" in some of the outdoor lines from the telephone company. If the problem persists, contact a service representative from the phone company and have the lines tested.

In instances where you are receiving bad interference from local radio transmissions (a nearby ham or CB radio operator, for example), simply install a .001 microfarad capacitor across the microphone unit in the telephone handset. The telephone companies have capacitors on hand for this purpose and will generally provide one free or at low cost.

Troubleshooting Chart

Possible Causes	Solution
Problem: Telephone is dead.	
Bad connector or outlet	Test; repair/replace
Telephone component is bad	Test; repair/replace
Incoming line is bad	After you've completed all of your own tests, contact the phone company
Problem: Intermittent operation; noise on the line.	
Bad connector or outlet	Test; repair/replace
Telephone component is bad	Test; repair/replace
Incoming line is shorted	Test; contact phone company
Incoming line is bad	After you've completed all of your own tests, contact the phone company
Interference	Install .001 capacitor

Toasters

The toaster is one appliance that almost *everyone* forgets to clean. Yet a crumb-clogged toaster is one of the major fire hazards in American homes today. If too many of those crumbs fall directly onto the surface of a red-hot heating element, they can easily erupt into flames. There is no excuse for having this danger in your home. Toasters are actually among the easiest of home appliances to maintain, and not really all that difficult to repair.

Most units have hinged crumb trays, which unsnap and swing outward from the bottom. (Always unplug a toaster from the wall outlet before cleaning.) If the trays are removable, wash them in the sink with your dishes and replace after thoroughly dry. If not removable, brush off crumbs or residue, or wipe the trays with a damp cloth. It is usually okay to use soap-filled commercial nylon or plastic scouring pads to remove any stubborn spots or stains. Dry thoroughly, however, before reinserting the tray.

Toasters usually consist of the cord and plug, at least one thermostat, heating elements (usually two per slice to be toasted), mechanical and spring-loaded parts, and a spring-loaded on/off switch. Some units have solenoids and auxillary heating coils around bimetallic sensors to signal when toasting is complete. Most of the components are easily tested for malfunction, and most of the components can be repaired or replaced with relative ease. The trick is to pay close attention to your disassembly procedure and to make sure that all the springs get back into their proper places.

With quite a few toasters, the heating elements are arranged as inside/outside and are not interchangeable.

Sophisticated models have temperature controls for toasting bread

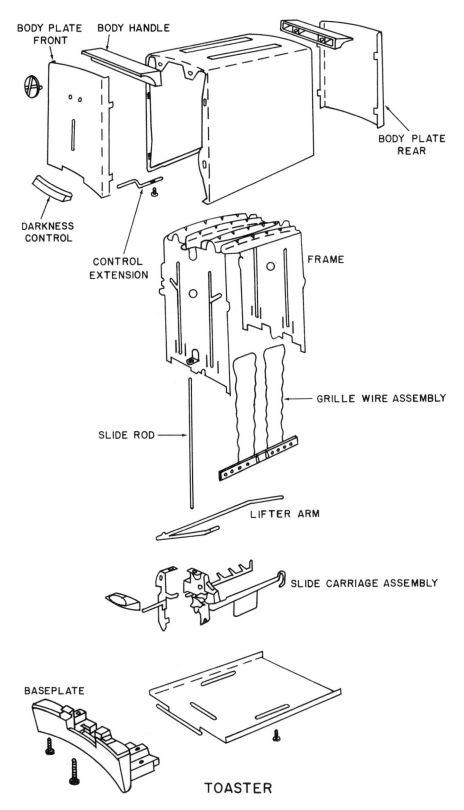

BODY PLATE
FRONT

BODY HANDLE

BODY PLATE
REAR

DARKNESS
CONTROL

CONTROL
EXTENSION

FRAME

GRILLE WIRE ASSEMBLY

SLIDE ROD

LIFTER ARM

SLIDE CARRIAGE ASSEMBLY

BASEPLATE

TOASTER

light, medium, or dark. When testing electrical malfunctions in the more expensive models of toasters, don't neglect to track down and check all safety switches and interlocks. They could be the hidden cause of an appliance breakdown.

The most frequent breakdowns occur due to worn or broken plugs or cords, or weakened or damaged springs. The heat inside the toaster can cause premature aging of the parts. As with other appliances that heat, the cause of a malfunction is often visible upon inspection.

A common problem is with the switch. As the switch is pushed down, the bread is lowered into the toaster and a latch holds it down until a thermostat or timer releases the catch and allows the toast to pop up. If the latch doesn't work, the toast won't stay down. If the thermostat, timer, or other automatic release fails, the toast might stay down to fill your kitchen with smoke.

If one side of the bread toasts and the other doesn't, the trouble is probably the heater element on that side. If both sides toast unevenly, the most likely cause of the problem is the bread itself. If the bread pops out before done, or burns, you've either set the selector incorrectly or the thermostat needs replacement.

DISASSEMBLY

The heating element is a series of wires or coils mounted on a frame with insulators. There are usually two elements for each slot, to toast both sides at the same time. These two are connected together so that both come on simultaneously; most of the time all the heating wires come on, regardless of how many slots are actually in use.

With most models, the heating element and holder slides or snaps out from below. In a few cases, all disassembly hardware is reached through the toaster top opening. In others, access is from a bottom plate or by way of a mounting panel at the back of the toaster. Regardless of the disassembly procedure, pay close attention to what you are doing. Most toasters have fingers that lift the toasted bread. These will often fall out along with the heater element.

For disassembly, you will usually have to remove all external knobs and controls. These usually pull off. Before you start yanking on the knobs, though, inspect them. You might find small holding screws that have to be loosened before the knob can be removed.

Carefully examine the case to see how it is all held together. Holding screws for the case are sometimes hidden beneath the trim plates. Other times it will all come apart by removing screws from the bottom. With still other toasters, the two sides snap off for disassembly.

You are likely to encounter push-button, rotary, and slide switches (lever switches) in a toaster. You will find that most are of the DPDT type. All can be easily tested with the VOM. While testing the switch, examine the related mechanical latches for any wear or damage.

There may also be small contacts inside. A visual examination will reveal if pitting or corrosion is causing a problem. A VOM test for continuity will help in your diagnosis.

With toaster-type devices, cleanliness is an essential safety precaution. The heating elements become red or even white-hot during cooking, and the presence of grease splatters or loose food particles creates a definite fire hazard. After every use of a toaster, let it cool and then turn it upside down and sideways, physically shaking out any food particles. Depending upon frequency of use, once monthly or once semi-annually, disassemble the unit more completely for a thorough cleaning of the interior.

All electrical appliances carry warnings that they should not be immersed in water. This is because moisture is an electrical conductor and can generate hazardous short circuits and because some moving parts such as motors and gears may be ruined if they become wet and are allowed to corrode. However, if one uses the proper precautions there are *some* electric heating elements that can be cleaned with water — if they are left open afterward in the direct sun or in a well-aired space in order to dry completely before reassembly and use. Heating elements in toasters are examples, if they can be easily removed and lightly washed as completely separate components. This can be helpful in instances where there has been a heavy buildup of debris. Be sure to allow the element to dry completely before putting the unit back together and turning it on.

If you are leary of using water, then use a commercial oven cleaning spray solution. Follow the manufacturer's directions, and use the solution to clean the interior of the toaster and the surface of the heating wires or coils on a regular basis. This is not only a wise fire-prevention maneuver, but something that will greatly extend the life of the appliance.

Troubleshooting Chart

Possible Causes	Solution

Problem: Toasting fails to begin; toaster does not operate; on lamp does not light.

Possible Causes	Solution
Faulty power cord or plug	Check cord and plug
Defective switch	Check switch
No power to outlet	Check outlet
Heating element(s) worn out	Test; replace

Problem: Breaker or fuse blows when unit is plugged in or turned on.

Possible Causes	Solution
Circuit overload	Try on another outlet
Short circuit	Test

Problem: Toast cooks but does not pop up, or won't stay down.

Possible Causes	Solution
Crumbs, etc. in mechanisms	Clean
Broken or damaged spring	Examine; repair or replace
Latch mechanism damaged or broken	Examine; repair or replace
Thermostat faulty	Test; replace

Trash Compactors

Modern trash compactors use a fairly large electric motor, geared down so that levers and push rods can force a large metal plate against trash poured into the container box. The motor action is slow speed, with either direct-drive gears or belt and pulley arrangements designed to squeeze household discards into small blocks for easier disposal.

Depending upon the model involved, the compactor may also bag the compressed trash into disposable heavy-duty garbage bags, may have some method for spraying the box contents with disinfectant or deodorizing chemicals, and may have a separate "crusher" to flatten cans before they are placed with other garbage for the final compacting action.

Consult the owner's manual before attempting any service or maintenance work on a home compactor. If you have lost the manual, write directly to the manufacturer to get a new copy. The manual is a valuable source of information on how to operate and maintain the appliance.

The simplest models have a motor and its associated mechanical drives, a power switch, and an electrical cord and plug. Most also have top and bottom limiting switches or relays that prevent the ram inside from traveling too far in either direction.

Complex models have speed and power gear changes, clutches, safety interlocks, switching networks for auxilliary services, and overload trips. A few even have automatic deodorizers that give a spritz to remove odors from the garbage.

Mechanical jamming leading to bent or broken parts is a major malfunction. Careful study of the device as it is taken apart (and before) will show you the proper way to replace any needed new parts. Your notes and sketches are critical for reassembly also.

A split phase motor is usually used to power the compactor. Chapter 5 gives you more information on this type of motor, as well as details on testing, servicing, and maintaining motors in general.

Maintenance on a compactor is usually limited to regular cleaning. You'll probably have to limit your lubrication efforts to those times when the compactor is open for other servicing.

For lubrication, keep in mind that some parts are under high pressure. Regular lubricating oil will not handle the job. A special high-pressure lubricant is needed, such as lithium grease. Your owner's manual will often specify what is needed and where.

While the appliance is working normally, pay close attention to the various noises it makes. Because of the overall noisiness of the compactor, it's not always easy to hear other sounds that would indicate the beginnings of a problem. By becoming consciously aware of what sounds it is *supposed* to make, recognizing other sounds will be easier.

DISASSEMBLY

Freestanding units are usually easier to open than built-in ones, since you can move the compactor away from walls and other objects for better access. The built-in unit may or may not have to be removed from its space in the cabinet before you can get inside.

Ideally, the unit you select for purchase should have appropriate access panels. Units that slide into a tight space between cabinets should have removable panels in the front. Most compactors are accessible only from the back or from the bottom.

Where possible, study instructions that come with your own particular make and model. Basic disassembly is often given in the owner's manual. During disassembly, make careful notes as to the positioning of all mechanical drive parts and all levers and pushrods, etc. In particular, pay attention to the placement of bushings, washers, and lock-nuts.

Proper action of the compactor after reassembly depends upon the correct positioning of many key components. Many push rods or levers are held in place with cotter pins or by shear pins. A shear pin is a rolled metal dowel inserted into a hole to connect two moving parts. It is designed to break, or shear off, in case of jams. This prevents the breakage of more important and more expensive parts. Replacement shear pins are less than 50¢.

If cotter or shear pins have to be bent or removed in order to disassemble the appliance, it is best to replace with new pins. Pins that have been bent for removal lose their tensile strength and are more liable to future breakage under normal use.

Troubleshooting Chart

Possible Causes	Solution
Problem: Machine fails to start, blows fuses/breakers; pilot light fails to light; on/off switch does not work.	
Wall outlet, power cord, plug, power switch, fuses/breaker, internal reset button, etc.	Check; repair or replace
Door or drawer not closed	Close it!
Bad safety switch	Test; replace
Problem: Voltage at switch ok, but motor does not energize.	
Faulty motor	Test; repair or replace
Bad control component	Test; replace
Problem: Appliance works, but difficult or impossible to change speeds.	
Speed switches	Test; repair or replace
Problem: Can hear motor turn on but nothing moves.	
Binding of some part	Disassemble; examine
Motor jammed or malfunctioning	Test; lubricate; repair or replace
Bad relay	Test; replace
Loose, slipping, or broken drive belt or gear	Examine; repair or replace
Problem: Appliance seems to be working, but compacting operation is not satisfactory.	
Improper trash loading	Check manual to see that right loading method is used.
Broken or bent parts	Examine; repair or replace
Binding parts	Examine; lubricate, repair or replace
Loose, slipping, or broken drive belt or gear	Examine; repair or replace

Vacuum Cleaners

There are two basic types of vacuum cleaner. A canister unit has a main body that houses the motor and other functional parts, with a hose attached. The bag is located inside the body. The upright vacuum cleaner has a housing at the bottom for the motor with an arm coming up out of the housing, to which the bag is attached.

The greatest enemy to the vacuum cleaner is dirt — the very reason that it exists. As the cleaner does its job, sucking up dirt and dust, that dirt can get into the working parts.

Since the bag for the canister-type is inside the body, along with the motor and other working parts, quite often there is a filter inside to help protect those parts. If this filter becomes dirty, it may cause the motor to overheat and can do permanent damage. It's important to clean the inside of the canister body thoroughly each time you change the bag.

Many upright units, and some canister-types, make use of an agitating brush to help the cleaning action. A drive belt or, more rarely, drive gears, are used to spin the brushes. These brushes are especially helpful when cleaning carpets or rugs. However, as they pick up hair, string and other debris, it's all too easy for the arm to bind. Meanwhile, the motor will keep trying to turn the arm and this friction can even melt the drive belt.

After each use, turn the vacuum cleaner over and remove any hair, string, or threads on the brushes. Occasionally remove the arm completely so that you can clean the dust from the bearings at the ends of the arm. While doing this, check the drive belt to be sure that it hasn't become worn or stretched.

It's also important to change the bag often. Most people let the bag fill completely before putting in a new one to "save money." It doesn't.

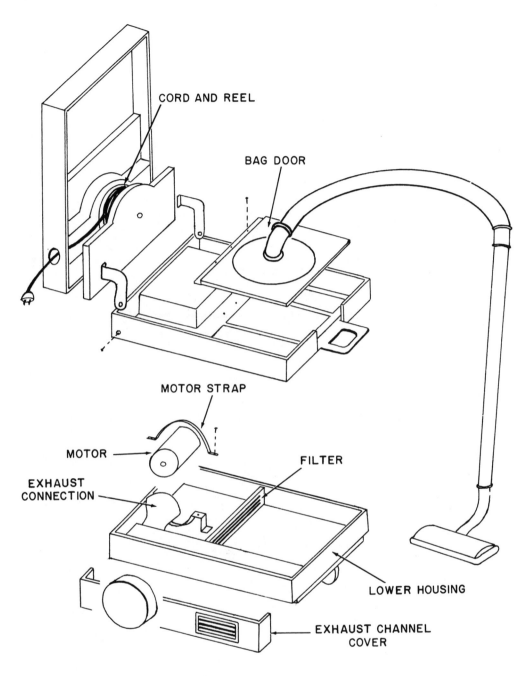

CORD AND REEL

BAG DOOR

MOTOR STRAP

MOTOR

FILTER

EXHAUST
CONNECTION

LOWER HOUSING

EXHAUST CHANNEL
COVER

CANISTER VACUUM CLEANER

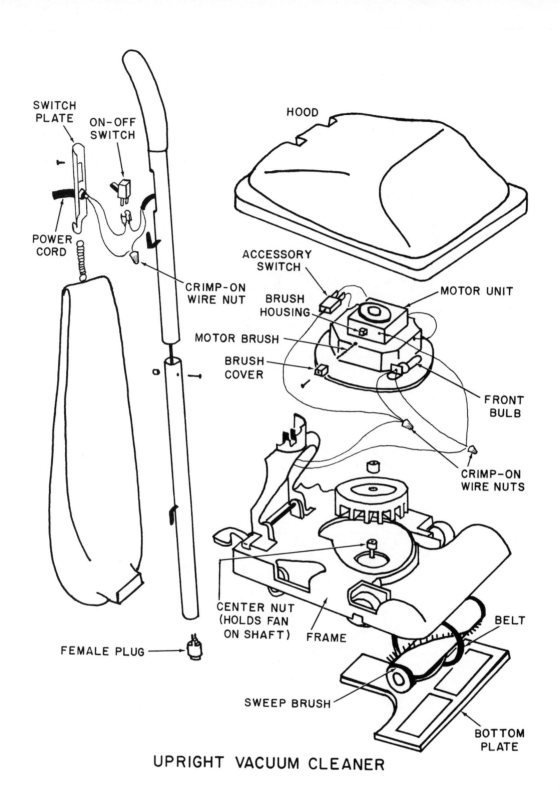

SWITCH PLATE

ON-OFF SWITCH

HOOD

POWER CORD

CRIMP-ON WIRE NUT

ACCESSORY SWITCH

BRUSH HOUSING

MOTOR UNIT

MOTOR BRUSH

BRUSH COVER

FRONT BULB

CRIMP-ON WIRE NUTS

CENTER NUT (HOLDS FAN ON SHAFT)

FRAME

BELT

FEMALE PLUG

SWEEP BRUSH

BOTTOM PLATE

UPRIGHT VACUUM CLEANER

As the bag becomes fuller, the efficiency of the vacuum cleaner decreases, and the strain on the unit increases.

Every unit will have at least one motor to drive fan blades that create the vacuuming action. This motor may also be coupled to the brush arm with a belt. A few units make use of a second motor for this function. The motor is usually a sealed unit to help keep the dust out. If the motor has to be replaced, the entire motor housing section almost always must be replaced. Breaking the seals to get at the motor is difficult (or impossible) and will usually cause trouble sooner or later.

Lack of suction should be tracked to its source. For a canister vacuum, check the amount of suction at the end of the hose; then remove the hose from the body of the cleaner and check the amount of suction there. If there is little suction at the nozzle, but plenty at the body, you know that the hose has to be repaired, or more likely, replaced. Doing the same with an upright cleaner isn't as easy, and is often impossible.

Troubleshooting Chart

Possible Causes	Solution
Problem: Machine fails to start, blows fuses/breakers, on/off switch does not work.	
Wall outlet, power cord, plug, power switch, fuses/breaker, internal reset button	Check; repair or replace
Bad safety switch	Test; replace
Problem: Voltage at switch ok, but motor does not energize.	
Faulty motor	Test; repair or replace
Bad switch	Test; replace
Problem: Appliance works, but difficult or impossible to change speeds.	
Speed switches	Test; repair or replace
Problem: Motor turns on, but nothing moves.	
Binding of some part	Disassemble; examine
Motor jammed or malfunctioning	Test; lubricate; repair or replace
Loose, slipping, or broken drive belt or gear	Examine; repair or replace
Agitator jammed	Examine; clean
Problem: Poor suction; poor cleaning action.	
Bag overfull or clogged	Replace
Filter clogged	Clean or replace
Hoses damaged; air leaks	Repair or replace hoses and gaskets
Agitator or brush jammed	Examine; clean
Drive belt or gear bad	Examine; repair or replace

Glossary

ac Alternating current. Supplied by power companies to homes. It is electricity moving back and forth along a wiring circuit, first in one direction and then in another.

amp Ampere. The unit for measuring how much electricity flows past a given point in one second of time when pushed by one volt of electrical pressure. It is somewhat similar to gallons per minute when measureing water flow through a pipe.

analog A signal that varies directly in proportion to another signal. An analog meter uses a needle that swings anywhere across a scale; compared to a digital meter which gives a reading in discrete numbers.

aquastat A device used to measure and control the temperature of the water in a hot water heater, or in a furnace that uses hot water.

armature The part of an electric motor that turns. The *armature winding* carries the current. Also the moving part of a relay.

armored cable More commonly called *BX*. Electrical wires encased in a flexible metallic sheath.

backflow Water moving in a direction other than normal, or from an external source other than the usual one.

battery A device that holds and/or generates dc power via a chemical reaction.

bearing A supporting part that holds a motor shaft and allows it to spin.

bibb The correct name for a faucet with threads. For example, an outdoor faucet that accepts a hose is correctly called a hose bibb.

bimetal Two metals binded together in a strip, with different bending temperatures. The two act as a temperature sensitive switch to control heating or cooling appliances.

breaker A type of "fuse" that does not melt, but that opens an electrical switch when an overload trips it. Somewhat similar to a switch. Can be reset.

brush A conductive "plug" that makes contact between the commutator of a motor and the source of power. The brush is usually made of graphite or carbon.

bushing An insulator to be placed around the wires extending from the end of a BX cable, to prevent the cut edges of the metal sheathing from damaging the insulation of the wires extending beyond.

BX More correctly called Type AC. A metal-armored cable for use in areas that are always dry. Often used in areas where Romex does not meet building codes.

cam An irregularly shaped device that when rotated on a shaft causes a switch or other mechanism to be tripped.

capacitor An electronic component that stores a charge. It also passes ac while blocking dc. Commonly used as a booster for motors and in power supplies.

centrifugal switch A switch that works as the speed of a motor reaches a certain level, usually to cut out the startup winding and to cut in the main winding.

check valve A valve that prevents water from flowing in the wrong direction. Sometimes incorrectly used to describe a safety valve that "lets go" when too much pressure builds up, such as in a water heater.

circuit A complete path for the flow of electricity, from the source (breaker box or wall outlet), through the appliance, and back to the source through the second connecting wire of the wall plug.

circuit breaker A modern replacement for old style fuses, this is a switching device designed to automatically turn off whenever a circuit overload or short circuit tries to draw more current than an electrical device is supposed to use. The circuit breaker is simply reset to return electricity to the circuit.

compressor A device used to compress a gas, such as freon. The expanding gas can then be used to cool.

commutator A part of the armature of a motor. Connects to the windings (coils) of the armature. A brush or set of brushes connected to the power supply makes contact with the commutator, and thus with the motor. Most commutators are made of pieces of copper on a central hub of steel.

condenser An old-fashioned term for a capacitor. More commonly, the part of a heating or cooling system that causes a gas to condense to a liquid.

conduit Metal or plastic tubing, used to run cables outdoors or underground, or in areas where exposed wires are in danger of fraying, cutting, or getting wet.

cps Cycles per second.

current The movement of electrical energy in a wire or circuit. Measured in amps.

dc Direct current. The type of electricity provided by batteries and needed for the proper operation of some small appliances. This is current that moves in one direction only, from the source, through the appliance, and back to the source. Some appliances designed to operate ac or dc have a built-in converter, or a power supply, to change ac from the power company into the dc needed to make the device function.

digital Signals with a discrete, on/off state, as opposed to analog signals, which have a continuously changing state.

distribution box Also called a *junction box*. A metal or plastic box, mounted in the wall, through which the wires carrying the incoming power enter the house. These wires are then connected to the wires that run to various outlets, switches, and appliances that operate from that particular circuit.

DPDT Double-pole, double throw switch.

field winding The coil of wire in a motor or other device that generates a magnetic field when current is applied.

fin comb Small pieces of metal in a pattern to disperse the heat generated by an appliance.

fuse A safety device to prevent electrical overloads, and to keep short circuits from causing wires to overheat to the point where they become fire hazards. Fuses are rated according to the amount of current (amps) they can carry, and when a circuit attempts to draw more current than the rated fuse value, it will burn out.

gauge Unit for measuring wire size. The lower the gauge number, the larger the wire, thus the greater current carrying capacity.

GFCI Ground fault circuit interrupter.

governor A device used to control the speed of a motor.

ground Actual earth, or the electrical ground or chassis point of an electrical circuit that provides the equivalent of earth. Sometimes used to refer to the return path of a circuit.

horsepower A measurement of physical work accomplished by an electrical device. When 746 watts of electrical power is used, it is equivalent to one horsepower.

hot wire The "high" side or power-carrying wire of a home wiring circuit, usually identified by a black or red colored insulation. Neutral or ground wires are usually white.

IC Integrated Circuit, or IC chip.

insulation A protective nonconducting coating of rubber, plastic, cotton, or lacquer, used to prevent electricity from jumping across one wire to another.

live circuit A circuit that is energized, or is carrying electrical current.

load What is being driven or powered by a device such as a motor. Also used to describe the amount of power used.

mercury switch A switch that uses a bubble of mercury to make or break contact when the containing tube is tipped.

microswitch A very small switch, or a switch with a very small contact for activation.

Ohm's law The mathematical formula relating voltage, current, and resistance. The three basic formulae are: $E = IR$, $I = E/R$, and $R = E/I$, where E is voltage, I is current, and R is resistance.

open circuit Sometimes called an *open*. A circuit with a break in a wire (like an open hole into which electricity disappears and does not pass through.

overload The condition when two many appliances or lights are placed on a single electrical circuit, exceeding the maximum amperage capable of being supplied by the circuit. Many household circuits are 15 amps.

polarizing Identification (usually by color code) of the hot or positive ($+$) side of a circuit or the ground or negative ($-$) side of the circuit. These codes help to prevent a repairman from making an accidental short circuit, connecting a hot wire directly to a ground wire.

polarized A plug or outlet that has a definite ground and hot, or positive and negative electrical or magnetic connections. In polarized plugs and outlets, one prong and one opening will be larger than the other.

power law The mathematical formula for power, in watts. Expressed as $P = IE$, where P is power in watts, I is current in amps, and E is voltage. If 10 amps of current is flowing with 117 volts, 1170 watts of power are being consumed.

relay A switch-like device. As current is applied, (e.g., when a built-in timer goes off), the relay causes an internal electromagnet to either make or break contact.

resistance The opposition to the flow of electrical current, measured in ohms.

rheostat An electronic component having a variable resistance. This regulates the amount of current, and thus the power being used. It is often used to control motor speed or temperature of a heating element.

Romex A cable covered with plastic or rubber rather than metal.

rpm Revolutions per minute; usually refers to motor speed.

schematic A line drawing showing the electrical connections between various circuits and devices.

service entrance The technical name for the main fuse box or breaker box, where power from the utility company is brought into the house. Total service amperage for most home wiring circuits is 120 to 160 amps, depending upon local housing codes.

short circuit An accidental connection between two points in an electrical circuit, when the hot leg of a circuit accidentally touches the ground leg (or your body) without going through an electric appliance. This provides a path for the maximum amount of current available from the power company, and trips any breakers or fuses. Short circuits can cause sparks and are fire dangers.

solder Generally, a mixture of tin and lead with a relatively low melting point. Used to provide a clean and corrosion-free joint.

solenoid A device in which an electromagnet pushes and pulls at a shaft, causing an inwards/outwards motion, and thus making or breaking mechanical contact. Often used as an electrically controlled valve.

SPST Single pole, single throw switch.

solderless connectors Crimp or twist caps to be placed over exposed wires, making a good electrical connection without solder.

test light A neon or low-wattage bulb with two test-probe leads. Can be used to check for the presence of voltage, ac or dc.

thermocouple A device that converts changes in temperature to electrical energy, often in the form of pulses to control other devices or equipment.

thermostat A device sensitive to temperature changes; it throws an internal switch when certain temperature levels are reached. Most are bimetallic, containing two metals that expand and contract at different rates and make or break electrical contact, depending on how far they are "bent" by temperature variance.

torque Measure of rotating force around an axis.

transformer A device used to change the value of the ac voltage going through it. The most common purpose in the home is to reduce line voltage to the needed value.

voltage The electrical "pressure" that moves electrical energy through a circuit. Measured in volts.

VOM   A volt-ohm-milliammeter, for measuring voltage, current, and resistance in electrical circuits.

watt   The unit of use of electrical power. One watt used for one hour is a watt-hour. 1,000 watt hours is a kilowatt hour, the measurement a utility company uses in order to bill you for electricity each month.

winding   The conductive wire of a motor that is wrapped and coupled inductively to the magnetic core.

Index

See also Glossary, pages 245-250

garage door openers, 170-175
grass trimmers, 176-181
ground wire, 46
grounded outlets, testing, 20
grounded wire, 47
grounding circuits, 45
grounding wire, 47

heat, and wire type, 44
heaters. *See* space heaters
heating elements, 14
  basic functioning of, 98-102
  in broiler ovens, 134-136
  cleaning of, 113
  as danger spot, 33
  in deep fat fryers, 154, 156
  diagnosis for, 102-104
  in electric blankets, 158
  in frypans, 215
  gauge of wire for, 45
  lug connections, 100-101
  maintenance of, 112-113
  plug-in units, 99-100
  repairing, 106-107
  resistance reading of, 70
  in space heaters, 223
  testing, 104-106
  in toasters, 234
  in water beds, 159
  welded or soldered units, 101-102
heating pads, 158-161
horsepower, 80
hot plates, 182-184
hot wire, 46
housing codes, 39
humidostats, 112
hysteresis/synchronous motors, 84

impedance, 105
inspection, after rewiring, 44
installation, 119-120
instruction manuals, 119
insulating tape, 15
insulation, color of, 45-46
interlocks, 134
irons, 185-188
  heating elements in, 105
  wire gauge for, 45

jewelry, and safety, 30
juicers, 189-191
junction boxes, 49

kitchen appliances, power cord size, 55
knives, electric, 162-165

labeling, of branch circuits, 39-40
labor charges, 2
lamps
  power cord size for, 55
  wire gauge for, 45
lawn mowers, 176-181
  drive belt replacement on, 96
leaf blowers, 180
light fixtures, short circuit in, 42
liquid porcelain, 122-123
local building authority, 43
lubrication, 89, 122
  for motors, 96
  for sewing machines, 204
  for trash compactors, 238

machine oil, 15
magnetic clutches, 86
magnetism, 80
maintenance
  general, 120-122
  preventive, 15, 114
mechanical malfunction, 70
  of motor shafts, 89
  in power tools, 193
metal-armored cable, 44
microchips, 75
microwave ovens
  voltage changes and, 50
  wire gauge for, 45
mixers, 127-132
moisture, and wire type, 44
motor circuit boards, testing, 87
motor shafts
  freedom of movement of, 88
  mechanical problems with, 89
motors, 79-97
  alternating current, 81-85
  for clocks, 146
  compound-wound, 81
  direct current, 85-86. *See also* direct
    current (dc) motors
  disassembly of, 95
  formula for speed of, 84
  hysteresis/synchronous, 84
  low voltage effect on, 50, 52
  maintenance of, 95-97
  mounts for, 87
  permanently sealed, lubricating,
    92-93

potential problems with, 69-70
for power tools, 194
problems with, 89-91
PSC, 82-83
for pumps, 201-202
replacement of, 95
series-wound, 81
shaded-pole, 81-82
for shavers, 209
shunt-wound, 81
split-phase, 82
starting capacitor for, 34
testing, 92-93
thermal overload circuit breaker on, 91
universal, 81
in vacuum cleaners, 243
multimeter. *See* VOM (volt-ohm-milliameter)

National Electrical Code, 43
National Fire Protection Association, Inc., 43
National Fire Underwriters Board, 45
needlenosed pliers, 12
neutral wire, 46
nichrome, 99
noise, from fans, 166
nut drivers, 12

one-hand rule, 30
open circuits, in motors, 89
operator's manual, 88
outlets, 49
    power to, 65
    replacement of, 52-53
    testing and replacing, 20, 50-53
ovens. *See* broiler ovens
overheating, 68
overloading, 67
    checking for, 41-43
    heating elements and, 103
    space heaters and, 221

parts
    buying, 77-78
    salvaged, 3
permanently sealed motors, lubricating, 92-93
Phillips screwdrivers, 11
pliers, 12
plugs, 62-64
polarity, of VOM leads, 22

portable heaters. *See* space heaters
power cords
    damaged, 23
    on deep fat fryers, 154
    sizes for, 55
    testing for short circuit in, 24-25
power supplies, 75-77
    built-in, 22
    capacitors as, 34
    testing, 77
power surge, 42
power tools, 192-199
    drive belt replacement on, 96
preventive maintenance, 114
    alcohol for, 15
proprietary information laws, 86
PSC motors, 82-83
pumps, 200-202

radio-controlled devices, for garage door openers, 173
radio frequency (RF) interference, 34
    in telephones, 230
radios
    power cord size for, 55
    wire gauge for, 45
razors, electric, wire gauge for, 45
rectifier, in power supply, 76
repairs, cost of, 2
replacement parts, buying, 77-78
replacement parts list, 88
resistance, testing for, 18, 23-26, 106
return wire, 46
rewiring, wiring codes and, 43
RF interference, 34
rheostats, 98, 99, 107
    in deep fat fryers, 156
    in electric blankets, 159
    problems with, 104
Romex insulation, 44-45

safety, 4, 28-36
    broiler ovens and, 133
    capacitors and, 85
    extension cords and, 55-56, 120
    during fan repair, 168
    in fuse replacement, 42-43
    garage door openers and, 173
    power tools and, 192-193
    toasters and, 232, 235
    and wire reconnection within appliances, 72
safety wire, 47

salvaged parts, 3
screw terminal, attaching wire to, 58
screwdrivers, 10-11
sealants, for steam irons, 187
semiconductors, for temperature control, 108
series-wound motors, 81
service entrance, 38-39
servicing, professional, 118
sewing machines, 203-207
shaded-pole motors, 81-82
shavers, 208-213
shear pins, 238
short circuits, 42, 65
   blackened fuse from, 41
   in hot plates, 182
   locating, 66-67
   motor malfunction from, 90
   in power cord, testing for, 24-25
   testing for, 26
   in wiring, 55
shunt-wound motors, 81
silver soldering, 102
single-pole single-throw wall switch, 48
skillets, 214-216
slow cookers, 156, 217-220
small appliances. *See* appliances
socket set, 12
solder sucker (desoldering tool), 13, 59
soldering, 53-54. *See also* unsoldering
   silver, 102
soldering iron, 13, 14
solderless screw cap, 58
solenoids, 198
solid-state electronics, 75, 86
   clocks with, 146-147
   thermostats with, 108
space heaters, 221-224
   heating elements in, 105
   low voltage effect on, 50, 52
speed control
   for blenders, mixers or processors, 131
   for fans, 168
splices, 56-62
split-phase motors, 82
spot welding, 102
spray contact cleaner, 112
spray lubricant, 15
start-up capacitors, 34
steam irons, 185, 187
storms, electrical, 42

stoves, power cord size for, 55
supplies, 15-17
switch boxes, 49
switches
   replacement of, in power tools, 194
   testing, 26
   in toasters, 234
   wall, 47-49

tape, electrical, 15, 61
taps, 56-62
telephone lines, problems with, 230
telephones, 225-231
televisions, power cord size for, 55
thermal overload circuit breaker, on motors, 91
thermistors, 108
thermostats, 99, 107-111
   cleaning of, 113
   in deep fat fryers, 156
   in electric blankets, 159
   in irons, 185
   maintenance for, 113
   problems with, 104
   solid-state, 108
   testing, 108-111
three-pronged plugs, 64
three-way wall switch, 49
timers, 111-112, 145-148
tinning process, 56-57, 61
toaster ovens. *See* broiler ovens
toasters, 232-236
   wire gauge for, 45
tools. *See also* power tools
   general purpose, 9-14
   and safety, 31
torque, 80, 81
Torx driver, 11
transformer, in power supply, 75-76
transistors
   in telephones, 229
   testing, 87
trash compactors, 237-239
troubleshooting charts
   for blenders, 132
   for broiler ovens, 136
   for can openers, 140
   for chain saws, 144
   for coffee makers, 152-153
   for deep fat fryers, 157
   for electric blankets, 161
   for electric knives, 165
   for fans, 169

for food processors, 132
for frypans, 216
for garage door openers, 174-175
for heating pads, 161
for hot plates, 184
for irons, 188
for juicers, 191
for lawn mowers and grass trimmers, 181
for mixers, 132
for power tools, 199
for pumps, 202
for sewing machines, 206-207
for shavers, 213
for slow cookers, 220
for space heaters, 224
for telephones, 230
for timers, 148
for toasters, 236
for trash compactors, 239
for vacuum cleaners, 244
for water beds, 161
Type AC cable, 44

universal motors, 81
unplugging, and safety, 32
unsoldering, 58-59

vacuum cleaners, 240-244
  drive belt replacement on, 96
  power cord size for, 55
vacuum pumps, 200-201
valves, on pumps, 201
varistors, 108
ventilation, 120
voltage
  checking, within appliances, 74

effect of changes in, on appliances, 50
loss of, and appliance functioning, 56
measurement of, 18, 30
VOM to test, 19-22
voltage dividers, 112
VOM (volt-ohm-milliameter), 14, 17-27
  for diagnosis, 73
  range of, 22
  testing cartridge fuses with, 41
  testing heating element with, 104-105
  testing motors with, 93-95
  testing outlets with, 50

wall outlet, testing, 20-21
wall switches, 47-49
warranties, 71, 118
water beds, 158-161
water heaters, testing thermostats on, 110-111
WD-40 (spray lubricant), 15
windings
  break in, 89-90
  continuity between, 93
wire
  color of, 45-46
  gauge of, 16-17, 45
  types of, 44-47
wire clippers, 12
wiring
  home, 43-49
  repair of, 54-64
  splices and taps for, 56-62
  for telephones, 228
wiring circuits, total service amperage for, 39
wiring codes, 43, 45
wrenches, 12